Safety Management:
A Guide for Facility Managers

Second Edition

Safety Management: A Guide for Facility Managers

Second Edition

Joseph F. Gustin

THE FAIRMONT PRESS, INC.

CRC Press
Taylor & Francis Group

Library of Congress Cataloging-in-Publication Data

Gustin, Joseph F., 1947-
Safety management : a guide for facility managers / Joseph F. Gustin. -- 2nd ed.
p. cm.
Includes bibliographical references and index.
ISBN 0-88173-331-8 (alk. paper) -- ISBN 0-88173-512-4 (electronic) -- ISBN 0-8247-5040-3 (Taylor & Francis Distributor : alk. paper)
1. Industrial safety--United States--Management. 2. Facility management--United States. I. Title.

T55.G795 2007
658.3'82--dc22

2007020689

Published by The Fairmont Press, Inc.
700 Indian Trail
Lilburn, GA 30047
tel: 770-925-9388; fax: 770-381-9865
http://www.fairmontpress.com

Distributed by Taylor & Francis Ltd.
6000 Broken Sound Parkway NW, Suite 300
Boca Raton, FL 33487, USA
E-mail: orders@crcpress.com

Distributed by Taylor & Francis Ltd.
23-25 Blades Court
Deodar Road
London SW15 2NU, UK
E-mail: uk.tandf@thomsonpublishingservices.co.uk

Printed in the United States of America
10 9 8 7 6 5 4 3 2 1

0-88173-331-8 (The Fairmont Press, Inc.)
0-8247-5040-3 (Taylor & Francis Ltd.)

Table of Contents

Preface

Safety is both a people and an environmental issue. It involves people and productivity and the costs of managing each. It involves coordinating policies and operations with industry standards and practices as well as with government regulations.

This new, 2nd edition of *Safety Management: A Guide For Facility Managers,* is written for facility managers and other professionals who want to provide a safe and accessible environment for employees, students, patients, customers and all other people who enter their facility.

The key to reducing safety and health hazards is an effective safety management program. In turn, the cornerstone of a successful safety management program is management commitment and employee involvement—forming a team that more and more often includes the facility manager as an integral player.

Like the first edition of this book, *Safety Management* focuses on managing the safety function. It provides the reader with a greater understanding of safety and health issues, including liability. It also enables the readers to better carry out their responsibilities in ensuring a safe working environment.

While the primary audience of *Safety Management: A Guide for Facility Managers, 2nd ed.,* is the facility manager, other professional including safety compliance officers, risk managers, human resources managers and supervisors at all levels can also benefit from its contents—as both a learning tool and as a reference.

Introduction

SAFETY: THE OPPORTUNITY AND THE CHALLENGE

Facility managers are the professionals most responsible for integrating people with their physical environment. As such, they often find themselves facing a myriad of complexities and challenges.

Among these complexities is the issue of safety, along with the concomitant challenges of accelerating regulatory activity, OSHA inspections, soaring workers' compensation costs, increased employee litigation, the Americans with Disabilities Act and violence in the workplace.

Each of these challenges requires greater effort on the part of employers in identifying, correcting and preventing safety and health hazards.

Awareness and proactive management can turn the challenges into opportunities—and a more productive workforce that works with management to ensure a safe working environment.

Safety Management: A Guide For Facility Managers, 2nd ed. addresses the complexities, opportunities and challenges of safety in detail, explaining how to systematically identify, evaluate and prevent general and specific hazards.

Each chapter of this new edition addresses a key area of safety management. Chapter 1 for example, provides a summary of the statistics that affect employers and the workforce. The most currently available data is presented and should be considered within the context of the shared responsibilities of employers and employees in preventing accidents and injuries.

Chapter 2 discusses the rights and responsibilities of both the employer and employee. Chapter 3 addresses the Whistleblower Act and its impact upon both employers and employees.

Other chapters talk about violence in the workplace, ADA compliance, conducting the safety audit, recordkeeping, safety inspections and change management. *Safety Management*, the 2nd edition covers in detail several major aspects of safety and health management that are either not well-covered, or simply not covered at all by many texts. Chapter 8, for example, discusses the written safety plan in detail and contains a sample written hazard communication program. Each element of the hazard communication program safety plan is explained, providing readers with the means to adapt the sample plan to their own particular organization.

At the core of this 2nd edition is the recognition that corporations and other organizations have a moral, ethical and legal responsibility to provide a safe environment for all people who enter their facilities.

Chapter 1

Understanding Safety As a Workforce Issue

The desire to provide a safe, productive work environment—while reducing liability and the hidden bottom line losses of downtime—is driving facility managers and building owners to establish comprehensive safety and health management programs.

Having such a program at the facility can ensure a safe environment where accidents and illnesses are minimized, and where such incidents are handled properly.

SAFETY: THE WORKFORCE ISSUE

Because safety in the workplace is such a complex issue, managing the safety function can be a challenging process. This complexity is rooted in the fact that safety is a condition of employment and is, therefore, subject to the full spectrum of employment law, including worker's compensation law.

As a condition of employment, safety is part and parcel of the implied contractual agreement between employer and employee. Like any other contract that is entered into in good faith, each party makes commitments and each party assumes responsibilities.

Certainly, neither employee nor employer assumes that either one of them will become another statistic in the occupational injury and illness records.

Data provided by the U.S. Department of Labor's Bureau of Labor Statistics for American employers indicate the scope of risk of workplace fatalities, as well as workplace injuries and illnesses. This data reveals that in 2005 a total of 5,702 fatal work injuries occurred in the United States. Other key findings for 2005 note that:

- Fatal work injuries among workers under 20 years of age were up about 18 percent from the 2004 figure to 166 cases.

- Fatal work injuries involving women in 2005 were down 3 percent to 402 cases—the lowest total ever recorded by the fatality census.

- Fatalities among agricultural workers were up 23 percent from 145 in 2004 to 178 in 2005.

- Fatal work injuries among Hispanic workers increased by 2 percent in 2004 to a new series high, though the fatality rate for Hispanic workers was lower.

- Fatal falls were lower by 7 percent after reaching a series high in 2004.

- While the number of fatal work injuries in private construction continued to be the most of any industry sector the number of fatalities was 4 percent lower in 2005 than in 2004.

- Fatal workplace injuries attributable to hurricanes accounted for 29 fatal work injuries in 2005, though this total may rise as additional cases are identified and verified.

Key findings for non-fatal workplace injuries/illnesses for 2005 include:

- Non-fatal workplace injuries/illnesses occurred at a rate of 4.6 cases per 100 equivalent full time workers in private industry.

- There were a total of 1.2 million injuries and illnesses requiring days away from work in 2005, relatively unchanged from 2004.

- Median days away from work—a key measure of the severity of the injury or illness—was 7 days in 2005, reflecting 2004 figures.

- 135.7 of these injuries/illnesses per 10,000 full-time equivalent workers were in private industry.

- More than 4 out of 10 injuries/illnesses were sprains or strains, most involving overexertion or falls.

- The rate resulted from a total of 4.2 million nonfatal injuries/illnesses in private industry workplaces during 2005, relatively unchanged compared to 2004, and a 2 percent increase in the number of hours worked. Offset by the increase in work hours, these numbers reflect a slight decline.

- Incidence rates for injuries/illnesses combined declined significantly in 2005 for most case types, with the exception of cases with days away from work (see Figure 1-1 and Table 1-1).

The BLS also reports that in 2005:

- Sprains and strains were the leading nature of injury/illness in every major industry sector. There was a decrease of 4 percent in these injuries from 2004, led by the manufacturing sector, which experienced an 8 percent decline. Sprains and strains declined by 7 percent in goods-producing injuries and by 3 percent in service providing injuries. Trade, transportation, and utilities reported 172,380 sprains and strains, 34 percent of the total in 2005.

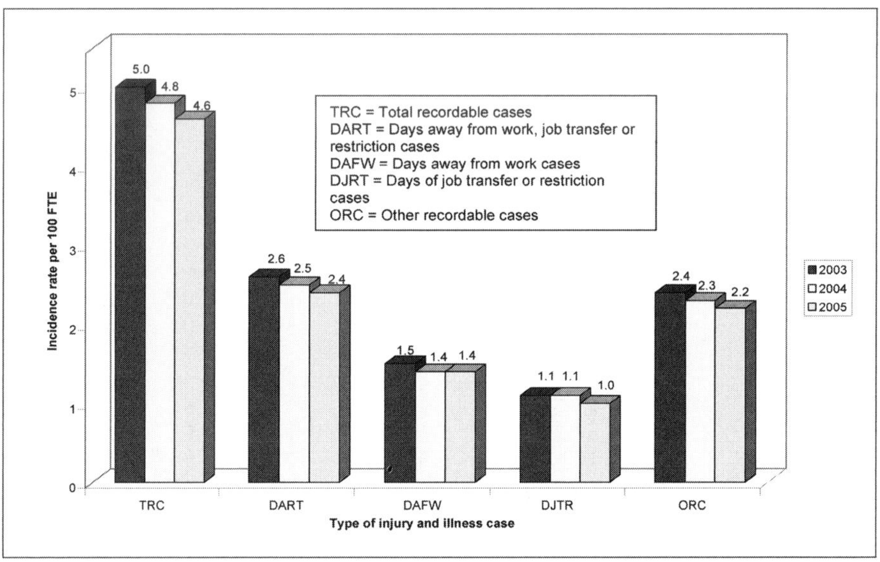

Figure 1-1. Nonfatal injury and illness incidence rates by case type, 2003-2005.

Table 1-1. Incidence rates[1] of nonfatal occupational injuries and illnesses by selected industries and case types, 2005

Industry[2]	NAICS code[3]	2005 Annual average employment[4] (thousands)	Total recordable cases	Cases with days away from work, job transfer, or restriction: Total	Cases with days away from work[5]	Cases with job transfer or restriction	Other recordable cases
Private industry[6]		109,127.0	4.6	2.4	1.4	1.0	2.2
Goods-producing[6]		22,897.1	6.2	3.4	1.8	1.6	2.8
Natural resources and mining[6,7]		1,517.7	5.1	2.9	1.8	1.0	2.2
Agriculture, forestry, fishing and hunting[6]	11	969.5	6.1	3.3	2.1	1.2	2.8
Crop production[6]	111	428.3	5.7	3.1	1.9	1.2	2.7
Animal production[6]	112	144.7	8.2	4.4	2.7	1.7	3.8
Forestry and logging	113	71.6	6.4	3.3	2.8	.5	3.1
Fishing, hunting and trapping	114	9.2	4.2	2.6	2.3	.3	1.6
Support activities for agriculture and forestry	115	315.7	5.4	3.0	1.9	1.1	2.4
Mining[7]	21	548.2	3.6	2.2	1.5	.8	1.4
Oil and gas extraction	211	124.4	2.1	1.2	.9	.3	.9
Mining (except oil and gas)[8]	212	209.2	4.1	2.7	2.0	.7	1.4
Support activities for mining	213	214.6	3.9	2.2	1.2	1.0	1.7
Construction		7,166.6	6.3	3.4	2.4	1.0	2.9
Construction	23	7,166.6	6.3	3.4	2.4	1.0	2.9
Construction of buildings	236	1,682.8	5.3	2.8	2.0	.8	2.5
Heavy and civil engineering construction	237	921.5	5.6	3.1	2.1	1.1	2.5
Specialty trade contractors	238	4,562.3	6.8	3.6	2.6	1.1	3.2
Manufacturing		14,212.8	6.3	3.5	1.5	2.0	2.8
Manufacturing	31-33	14,212.8	6.3	3.5	1.5	2.0	2.8
Food manufacturing	311	1,479.2	7.7	4.9	1.6	3.4	2.8
Beverage and tobacco product manufacturing	312	192.0	8.1	5.4	2.5	2.9	2.7
Textile mills	313	222.4	4.4	2.3	.8	1.4	2.1
Textile product mills	314	171.2	4.9	2.8	.9	1.9	2.0
Apparel manufacturing	315	263.7	3.3	1.5	.8	.7	1.8
Leather and allied product manufacturing	316	40.0	6.6	3.8	1.7	2.2	2.8
Wood product manufacturing	321	556.8	9.4	5.2	2.5	2.7	4.2
Paper manufacturing	322	486.2	4.4	2.5	1.2	1.3	1.9
Printing and related support activities	323	648.3	4.1	2.3	1.1	1.2	1.8
Petroleum and coal products manufacturing	324	112.3	3.0	1.8	1.0	.8	1.2
Chemical manufacturing	325	874.8	3.2	1.8	.8	1.0	1.4

See footnotes at end of table.

Table 1-1. Incidence rates[1] of nonfatal occupational injuries and illnesses by selected industries and case types, 2005 (*Continued*)

Industry[2]	NAICS code[3]	2005 Annual average employment[4] (thousands)	Total recordable cases	Cases with days away from work, job transfer, or restriction			Other recordable cases
				Tota	Cases with days away from work[5]	Cases with job transfer or restriction	
Plastics and rubber products manufacturing	326	800.8	7.1	4.1	1.7	2.4	3.0
Nonmetallic mineral product manufacturing	327	503.0	8.0	4.9	2.5	2.3	3.2
Primary metal manufacturing	331	466.1	9.1	4.7	2.2	2.6	4.4
Fabricated metal product manufacturing	332	1,509.5	8.0	3.8	1.9	1.9	4.2
Machinery manufacturing	333	1,153.7	6.5	3.0	1.4	1.6	3.5
Computer and electronic product manufacturing	334	1,310.7	2.0	1.0	.5	.6	1.0
Electrical equipment, appliance, and component manufacturing	335	436.9	5.2	2.8	1.1	1.7	2.5
Transportation equipment manufacturing	336	1,770.1	8.3	4.6	1.7	2.9	3.7
Furniture and related product manufacturing	337	565.7	7.3	4.0	1.7	2.3	3.3
Miscellaneous manufacturing	339	649.3	4.4	2.4	1.0	1.4	2.0
Service-providing		86,229.9	4.1	2.1	1.2	.9	2.0
Trade, transportation, and utilities[9]		25,564.9	5.2	3.0	1.7	1.3	2.2
Wholesale trade	42	5,724.8	4.5	2.7	1.5	1.2	1.8
Merchant wholesalers, durable goods	423	2,981.1	4.1	2.2	1.3	.9	1.9
Merchant wholesalers, nondurable goods	424	2,012.4	5.7	3.8	1.9	1.9	1.9
Wholesale electronic markets and agents and brokers	425	731.3	2.5	1.5	.8	.6	1.1
Retail trade	44-45	15,209.1	5.0	2.6	1.5	1.2	2.4
Motor vehicle and parts dealers	441	1,911.9	4.8	2.1	1.4	.7	2.7
Furniture and home furnishings stores	442	572.9	5.0	2.8	1.8	1.0	2.2
Electronics and appliance stores	443	531.0	2.4	1.3	.7	.7	1.1
Building material and garden equipment and supplies dealers	444	1,265.9	7.5	3.9	2.1	1.8	3.6
Food and beverage stores	445	2,809.0	6.0	3.2	1.9	1.3	2.8
Health and personal care stores	446	945.5	2.4	1.2	.8	.4	1.2
Gasoline stations	447	866.5	3.5	1.7	1.2	.5	1.8
Clothing and clothing accessories stores	448	1,404.6	2.7	1.0	.6	.4	1.6
Sporting goods, hobby, book, and music stores	451	649.0	3.3	1.3	.8	.5	2.0
General merchandise stores	452	2,918.9	6.7	3.9	1.7	2.2	2.9
Miscellaneous store retailers	453	908.5	3.6	2.0	1.2	.9	1.6
Nonstore retailers	454	425.5	4.6	2.7	1.3	1.3	1.9
Transportation and warehousing[9]	48-49	4,077.7	7.0	4.6	2.9	1.7	2.4
Air transportation	481	505.0	9.9	7.5	5.3	2.2	2.4
Rail transportation[9]	482	–	2.5	1.9	1.6	.2	.6
Water transportation	483	58.0	3.9	2.4	1.8	.6	1.5
Truck transportation	484	1,381.6	6.1	3.9	2.9	1.0	2.2

See footnotes at end of table.

Table 1-1. Incidence rates[1] of nonfatal occupational injuries and illnesses by selected industries and case types, 2005 (*Continued*)

Industry[2]	NAICS code[3]	2005 Annual average employment[4] (thousands)	Total recordable cases	Cases with days away from work, job transfer, or restriction: Total	Cases with days away from work, job transfer, or restriction: Cases with days away from work[5]	Cases with days away from work, job transfer, or restriction: Cases with job transfer or restriction	Other recordable cases
Transit and ground passenger transportation	485	384.3	6.2	3.5	2.3	1.2	2.7
Pipeline transportation	486	37.7	2.1	.9	.6	.3	1.2
Scenic and sightseeing transportation	487	27.6	5.3	2.4	2.1	.3	2.9
Support activities for transportation	488	546.8	5.5	3.4	2.3	1.1	2.2
Couriers and messengers	492	559.1	11.6	8.0	4.0	4.0	3.6
Warehousing and storage	493	577.3	8.2	5.4	2.2	3.2	2.8
Utilities	22	553.3	4.6	2.4	1.3	1.0	2.3
Utilities	221	553.3	4.6	2.4	1.3	1.0	2.3
Information		3,064.0	2.1	1.1	.7	.4	1.0
Information	51	3,064.0	2.1	1.1	.7	.4	1.0
Publishing industries (except Internet)	511	904.9	2.2	1.1	.7	.4	1.1
Broadcasting (except Internet)	515	324.9	2.0	1.0	.6	.4	1.0
Internet publishing and broadcasting	516	30.7	.2	(10)	–	–	.1
Telecommunications	517	998.7	2.6	1.5	1.1	.4	1.1
Internet service providers, web search portals, and data processing services	518	380.3	1.1	.5	.3	.2	.6
Other information services	519	49.6	1.2	.5	.3	.2	.7
Financial activities		7,994.2	1.7	.8	.5	.2	.9
Finance and insurance	52	5,879.7	1.0	.4	.3	.1	.6
Monetary authorities - central bank	521	20.8	1.8	.9	.5	.4	.9
Credit intermediation and related activities	522	2,852.1	1.0	.4	.3	.1	.7
Securities, commodity contracts, and other financial investments and related activities	523	786.7	.3	.1	.1	(11)	.2
Insurance carriers and related activities	524	2,132.4	1.2	.4	.3	.1	.8
Funds, trusts, and other financial vehicles	525	87.7	1.1	.4	.2	.2	.8
Real estate and rental and leasing	53	2,114.5	3.7	2.1	1.3	.7	1.6
Real estate	531	1,443.3	3.3	1.8	1.3	.5	1.5
Rental and leasing services	532	645.1	4.6	2.7	1.5	1.2	1.9
Lessors of nonfinancial intangible assets (except copyrighted works)	533	26.1	.9	.5	.5	(10)	.4
Professional and business services		16,732.9	2.4	1.2	.7	.5	1.2
Professional, scientific, and technical services	54	6,979.3	1.4	.6	.4	.2	.8

See footnotes at end of table.

Table 1-1. Incidence rates[1] of nonfatal occupational injuries and illnesses by selected industries and case types, 2005 (*Continued*)

Industry[2]	NAICS code[3]	2005 Annual average employment[4] (thousands)	Total recordable cases	Cases with days away from work, job transfer, or restriction: Total	Cases with days away from work, job transfer, or restriction: Cases with days away from work[5]	Cases with days away from work, job transfer, or restriction: Cases with job transfer or restriction	Other recordable cases
Professional, scientific, and technical services	541	6,979.3	1.4	0.6	0.4	0.2	0.8
Management of companies and enterprises	55	1,730.9	2.4	1.3	.6	.7	1.1
Administrative and support and waste management and remediation services	56	8,022.7	3.7	2.0	1.3	.7	1.7
Administrative and support services	561	7,688.3	3.4	1.8	1.2	.6	1.6
Waste management and remediation services	562	334.4	7.1	4.7	2.7	2.0	2.4
Education and health services		16,385.7	5.5	2.6	1.5	1.1	2.9
Educational services	61	2,129.3	2.4	1.0	.7	.3	1.5
Educational services	611	2,129.3	2.4	1.0	.7	.3	1.5
Health care and social assistance	62	14,256.4	5.9	2.8	1.6	1.3	3.1
Ambulatory health care services	621	5,066.2	2.8	1.0	.6	.3	1.8
Hospitals	622	4,286.6	8.1	3.3	1.8	1.5	4.8
Nursing and residential care facilities	623	2,835.6	9.1	5.7	2.9	2.8	3.5
Social assistance	624	2,068.1	4.3	2.2	1.3	.9	2.1
Leisure and hospitality		12,687.5	4.7	1.8	1.1	.7	2.9
Arts, entertainment, and recreation	71	1,865.5	6.1	2.9	1.6	1.4	3.2
Performing arts, spectator sports, and related industries	711	379.3	8.3	2.8	1.8	–	–
Museums, historical sites, and similar institutions	712	118.4	5.2	2.0	1.2	.8	3.2
Amusement, gambling, and recreation industries	713	1,367.8	5.6	3.0	1.5	1.5	2.6
Accommodation and food services	72	10,822.0	4.5	1.7	1.0	.6	2.8
Accommodation	721	1,806.8	6.1	3.3	1.7	1.6	2.8
Food services and drinking places	722	9,015.3	4.1	1.3	.9	.4	2.8
Other services		3,800.7	3.2	1.5	1.0	.5	1.7
Other services, except public administration	81	3,800.7	3.2	1.5	1.0	.5	1.7
Repair and maintenance	811	1,231.8	4.0	1.8	1.3	.6	2.1

See footnotes at end of table.

Table 1-1. Incidence rates[1] of nonfatal occupational injuries and illnesses by selected industries and case types, 2005 (*Continued*)

Industry[2]	NAICS code[3]	2005 Annual average employment[4] (thousands)	Total recordable cases	Cases with days away from work, job transfer, or restriction			Other recordable cases
				Total	Cases with days away from work[5]	Cases with job transfer or restriction	
Personal and laundry services	812	1,271.6	2.7	1.6	0.8	0.7	1.2
Religious, grantmaking, civic, professional, and similar organizations	813	1,297.3	2.7	1.1	.7	.3	1.7

[1] The incidence rates represent the number of injuries and illnesses per 100 full-time workers and were calculated as: (N/EH) x 200,000, where

N = number of injuries and illnesses
EH = total hours worked by all employees during the calendar year
200,000 = base for 100 equivalent full-time workers (working 40 hours per week, 50 weeks per year)

[2] Totals include data for industries not shown separately.
[3] *North American Industry Classification System* — United States, 2002
[4] Employment is expressed as an annual average and is derived primarily from the BLS-Quarterly Census of Employment and Wages (QCEW) program.
[5] Days-away-from-work cases include those that result in days away from work with or without job transfer or restriction.
[6] Excludes farms with fewer than 11 employees.
[7] Data for Mining (Sector 21 in the *North American Industry Classification System*-- United States, 2002) include establishments not governed by the Mine Safety and Health Administration rules and reporting, such as those in Oil and Gas Extraction and related support activities. Data for mining operators in coal, metal, and nonmetal mining are provided to BLS by the Mine Safety and Health Administration, U.S. Department of Labor. Independent mining contractors are excluded from the coal, metal, and nonmetal mining industries. These data do not reflect the changes the Occupational Safety and Health Administration made to its recordkeeping requirements effective January 1, 2002; therefore, estimates for these industries are not comparable to estimates in other industries.
[8] Data for mining operators in this industry are provided to BLS by the Mine Safety and Health Administration, U.S. Department of Labor. Independent mining contractors are excluded. These data do not reflect the changes the Occupational Safety and Health Administration made to its recordkeeping requirements effective January 1, 2002; therefore, estimates for these industries are not comparable to estimates in other industries.
[9] Data for employers in railroad transportation are provided to BLS by the Federal Railroad Administration, U.S. Department of Transportation.
[10] Fewer than 15 cases.
[11] Incidence rate less than 0.05.

NOTE: Because of rounding, components may not add to totals. Dash indicates data not available.
SOURCE: Bureau of Labor Statistics, U.S. Department of Labor

- The incidence rate for carpal tunnel syndrome decreased by nearly 14 percent.
- The part of the body most affected by work incidents was the trunk, including the shoulder and back, which accounted for 35 percent of all cases. Overall injuries to the trunk decreased by 4 percent from 2004.
- Of the trunk injuries/illnesses 63 percent involved back injuries.
- Floors, walkways, and ground surfaces accounted for 19 percent of all sources of injury/illness. Worker motion or position accounted for 15 percent.
- Assaults and violent acts decreased by 18 percent. The majority of these acts, almost two-thirds, occurred in health care and social assistance sector.
- Repetitive motion injuries/illnesses decreased by 10 percent.
- Falls from a ladder increased by almost 10 percent.

BLS also collects the time of day and day of the week the injury/illness occurred and the time the employee had spent on the job before the incident:

- Of the injuries/illnesses with days away from work which the time of the incident was reported, the four hours from 8:00 A.M. to noon accounted for 36 percent of the cases. The hours from noon to 4:00 P.M. accounted for 28 percent.
- In those cases where employers reported how long the employee had been on the job before the incident occurred, workers on the job from two to four hours incurred 27 percent of injuries/illnesses with days away from work in 2005. Employees on the job for more than eight hours accounted for 28 percent.
- Eighty-seven percent of injuries/illnesses occurred on Monday through Friday. The exception was the leisure/hospitality sector, were 16 percent of injuries/illnesses occurred on Saturday.

Industries Affected

Additional data show that:

- Goods producing industries as a whole had an injury/illness incidence rate of 6.2 cases per 100 equivalent full-time workers, while service-providing industries had a rate of 4.1 cases. The incidence rate for goods-producing industries declined by 0.3 cases and the rate for service-providing industries fell by 0.1 case per 100 equivalent full-time workers compared to the rates reported for 2004. Among goods-producing industry sectors, incidence rates during 2005 ranged from 3.6 cases per 100 full-time workers in mining to 6.3 cases in construction and in manufacturing (see Table 1-1). While rates among service-providing industry sectors ranged up to 7.0 cases per 100 full-time workers in transportation and warehousing, finance and insurance had the lowest rate within this domain at 1.0 case. Despite this low rate, finance and insurance was the only industry sector to experience a statistically significant increase in the overall incidence rate in 2005, rising by 0.1 case per 100 full-time workers.

- Small establishments (those employing 1-10 workers) reported the lowest rate for injuries/illnesses combined (2.0 cases per 100 full-time workers), while mid-size establishments (those employing 50-249 workers) reported the highest rate (5.8 cases) (see Table 1-2). While incidence rates remained relatively unchanged for establishments employing fewer than 1,000 workers, the rate for large establishments (those employing 1,000 or more workers) declined significantly in 2005 to 5.2 cases per 100 full-time workers, down from 5.4 in 2004.

- Fourteen industries, each having at least 100,000 injuries/illnesses combined, accounted for slightly more than 1.9 million cases (46 percent) of the 4.2 million total cases (see Table 1-3). These same fourteen industries also reported having at least 100,000 injuries/illnesses in both of the previous two years. Hospitals topped this group of industries in each of the last three years, followed by nursing and residential care facilities.

- Approximately 2.2 million injuries/illnesses were cases with days away from work, job transfer, or restriction; that is, they required

Table 1-2. Incidence rates[1] of nonfatal occupational injuries and illnesses by major industry sector and employment size, 2005

Industry sector	All establish-ments	Establishment employment size (workers)				
		1 to 10	11 to 49	50 to 249	250 to 999	1,000 or more
Private industry[2]	4.6	2.0	4.1	5.8	5.2	5.2
Goods-producing[2]	6.2	3.7	6.5	7.2	6.0	5.5
Natural resources and mining[2,3]	5.1	2.8	5.4	5.3	5.7	2.9
Construction	6.3	4.2	6.7	7.4	6.5	5.2
Manufacturing	6.3	2.8	6.6	7.3	5.9	5.5
Service-providing	4.1	1.6	3.4	5.3	4.9	5.0
Trade, transportation, and utilities[4]	5.2	2.2	4.4	6.5	6.7	6.4
Information	2.1	–	1.9	2.7	2.1	1.8
Financial activities	1.7	1.5	1.7	2.1	1.8	.9
Professional and business services	2.4	1.0	2.7	2.9	2.5	–
Education and health services	5.5	1.0	3.0	6.9	6.6	7.1
Leisure and hospitality	4.7	1.6	3.7	6.1	6.6	7.1
Other services, except public administration	3.2	1.8	3.6	4.9	3.6	4.9

[1] The incidence rates represent the number of injuries and illnesses per 100 full-time workers and were calculated as: (N/EH) x 200,000, where

N = number of injuries and illnesses
EH = total hours worked by all employees during the calendar year
200,000 = base for 100 equivalent full-time workers (working 40 hours per week, 50 weeks per year)

[2] Excludes farms with fewer than 11 employees.

[3] Data for Mining (Sector 21 in the *North American Industry Classification System*-- United States, 2002) include establishments not governed by the Mine Safety and Health Administration rules and reporting, such as those in Oil and Gas Extraction and related support activities. Data for mining operators in coal, metal, and nonmetal mining are provided to BLS by the Mine Safety and Health Administration, U.S. Department of Labor. Independent mining contractors are excluded from the coal, metal, and nonmetal mining industries. These data do not reflect the changes the Occupational Safety and Health Administration made to its recordkeeping requirements effective January 1, 2002; therefore, estimates for these industries are not comparable to estimates in other industries.

[4] Data for employers in railroad transportation are provided to BLS by the Federal Railroad Administration, U.S. Department of Transportation.

NOTE: Dash indicates data not available.
SOURCE: Bureau of Labor Statistics, U.S. Department of Labor

Table 1-3. Numbers of cases and incidence rates[1] of nonfatal occupational injuries and illnesses with 100,000 or more cases, 2005 (*Continued*)

Industry[2]	NAICS code[3]	2005 Annual average employment[4] (thousands)	Total cases (thousands)	Incidence rate
Hospitals	622	4,286.6	281.5	8.1
Nursing and residential care facilities	623	2,835.6	209.1	9.1
General merchandise stores	452	2,918.9	147.2	6.7
Transportation equipment manufacturing	336	1,770.1	146.8	8.3
Administrative and support services	561	7,688.3	141.1	3.4
Fabricated metal product manufacturing	332	1,509.5	121.8	8.0
Merchant wholesalers, durable goods	423	2,981.1	119.5	4.1
Building equipment contractors	2382	1,890.8	117.8	6.7
Food manufacturing	311	1,479.2	114.2	7.7
Full-service restaurants	7221	4,270.4	111.7	3.9
Supermarkets and other grocery (except convenience) stores	44511	2,290.5	110.7	6.5
Ambulatory health care services	621	5,066.2	110.6	2.8
Merchant wholesalers, nondurable goods	424	2,012.4	110.0	5.7
Limited-service eating places	7222	3,849.4	103.3	4.1
Private industry[5]		109,127.0	4,214.2	4.6

[1] The incidence rates represent the number of injuries and illnesses per 100 full-time workers and were calculated as: (N/EH) x 200,000, where

N = number of injuries and illnesses
EH = total hours worked by all employees during the calendar year
200,000 = base for 100 equivalent full-time workers (working 40 hours per week, 50 weeks per year)

[2] Totals include data for industries not shown separately.
[3] *North American Industry Classification System* — United States, 2002
[4] Employment is expressed as an annual average and is derived primarily from the BLS-Quarterly Census of Employment and Wages (QCEW) program.
[5] Excludes farms with fewer than 11 employees.

SOURCE: Bureau of Labor Statistics, U.S. Department of Labor

recuperation away from work, transfer to another job, restricted duties at work, or a combination of these actions. The remaining 2.0 million injuries and illnesses were other recordable cases that did not result in time away from work (see Table 1-4). The incidence rate for cases with days away from work, job transfer, or restriction was 2.4 cases per 100 workers, and the rate for other recordable cases was 2.2. Both of these rates decreased by 0.1 case per 100 equivalent full-time workers from 2004.

- Cases with days away from work, job transfer, or restriction are comprised of two case types; those requiring at least one day away from work, with or without job transfer or restriction, and those requiring only job transfer or restriction. The latter case type may involve shortened work hours, a temporary job change, or temporary restrictions on a worker's regular duties; for example, no heavy lifting. Separately, the rate for cases with days away from work restriction was 1.0 case per 100 workers (down from 1.1 cases in 2004). The rate in manufacturing for cases with job transfer or restriction (2.0) was higher than the rate for days-away-from-work cases (1.5) (see Figure 1-2). Among the remaining industry sectors included in this chart, the rate for cases with days away from work was higher than the rate for cases with job transfer or restriction.

Injuries and Illnesses

- **Injuries**—Of the 4.2 million nonfatal occupational injuries/illnesses in 2005, approximately 4.0 million, or 94.2 percent, were injures. Of these 4.0 million injuries, 2.7 million, or 68 percent, occurred in service-providing industries which employee 79 percent of the private sector workforce covered by this survey. The remaining 1.3 million injuries, 32 percent, occurred in goods-producing industries which account for only 21 percent of the private sector employment. (The Bureau of Labor Statistics, Quarterly Census of Employment and Wages provided the data.) The largest shares of injuries were in the manufacturing sector 20 percent, the health care and social assistance sector 16 percent, and the retail trade sector 15 percent (see Table 1-5 and Figure 1-2).

Table 1-4. Numbers of nonfatal occupational injuries and illnesses by selected industries and case types, 2005.

Industry[1]	NAICS code[2]	2005 Annual average employment[3]	Total recordable cases	Cases with days away from work, job transfer, or restriction			Other recordable cases
				Total	Cases with days away from work[4]	Cases with job transfer or restriction	
Private industry[5]		109,127.0	4,214.2	2,184.8	1,234.7	950.1	2,029.4
Goods-producing[5]		22,897.1	1,387.6	756.4	394.1	362.3	631.2
Natural resources and mining[5,6]		1,517.7	76.9	43.2	27.9	15.3	33.7
Agriculture, forestry, fishing and hunting[5]	11	969.5	54.6	29.5	18.9	10.6	25.1
Crop production[5]	111	428.3	23.3	12.5	7.7	4.8	10.9
Animal production[5]	112	144.7	13.1	7.0	4.3	2.8	6.1
Forestry and logging	113	71.6	4.1	2.1	1.8	.3	2.0
Fishing, hunting and trapping	114	9.2	.3	.2	.2	([7])	.1
Support activities for agriculture and forestry	115	315.7	13.7	7.7	4.9	2.7	6.1
Mining[6]	21	548.2	22.2	13.7	9.0	4.6	8.6
Oil and gas extraction	211	124.4	2.6	1.5	1.1	.4	1.1
Mining (except oil and gas)[8]	212	209.2	10.1	6.8	5.0	1.8	3.4
Support activities for mining	213	214.6	9.6	5.5	3.0	2.5	4.1
Construction		7,166.6	414.9	222.5	157.1	65.4	192.4
Construction	23	7,166.6	414.9	222.5	157.1	65.4	192.4
Construction of buildings	236	1,682.8	81.3	43.4	31.2	12.2	37.9
Heavy and civil engineering construction	237	921.5	51.0	28.5	18.9	9.7	22.5
Specialty trade contractors	238	4,562.3	282.5	150.5	107.0	43.5	132.0
Manufacturing		14,212.8	895.9	490.8	209.1	281.7	405.0
Manufacturing	31-33	14,212.8	895.9	490.8	209.1	281.7	405.0
Food manufacturing	311	1,479.2	114.2	73.2	23.5	49.7	41.0
Beverage and tobacco product manufacturing	312	192.0	15.4	10.2	4.7	5.5	5.2
Textile mills	313	222.4	9.6	4.9	1.8	3.2	4.6
Textile product mills	314	171.2	8.0	4.7	1.5	3.1	3.3
Apparel manufacturing	315	263.7	7.9	3.7	1.9	1.8	4.2
Leather and allied product manufacturing	316	40.0	2.5	1.4	.6	.8	1.0
Wood product manufacturing	321	556.8	52.3	28.8	13.7	15.1	23.5
Paper manufacturing	322	486.2	22.3	12.8	6.1	6.7	9.5
Printing and related support activities	323	648.3	25.6	14.3	7.1	7.2	11.3

See footnotes at end of table.

Table 1-4. Numbers of nonfatal occupational injuries and illnesses by selected industries and case types, 2005. (*Continued*)

Industry[1]	NAICS code[2]	2005 Annual average employment[3]	Total recordable cases	Cases with days away from work, job transfer, or restriction			Other recordable cases
				Total	Cases with days away from work[4]	Cases with job transfer or restriction	
Petroleum and coal products manufacturing	324	112.3	3.5	2.1	1.2	0.9	1.4
Chemical manufacturing	325	874.8	27.8	15.8	7.2	8.6	12.0
Plastics and rubber products manufacturing	326	800.8	58.4	33.7	14.0	19.6	24.7
Nonmetallic mineral product manufacturing	327	503.0	41.6	25.2	13.1	12.1	16.4
Primary metal manufacturing	331	466.1	45.0	23.2	10.6	12.6	21.7
Fabricated metal product manufacturing	332	1,509.5	121.8	58.2	29.2	29.0	63.6
Machinery manufacturing	333	1,153.7	75.8	35.1	16.4	18.7	40.7
Computer and electronic product manufacturing	334	1,310.7	26.4	13.3	6.1	7.2	13.1
Electrical equipment, appliance, and component manufacturing	335	436.9	22.9	12.2	4.7	7.5	10.8
Transportation equipment manufacturing	336	1,770.1	146.8	81.0	29.8	51.2	65.9
Furniture and related product manufacturing	337	565.7	40.8	22.3	9.5	12.9	18.5
Miscellaneous manufacturing	339	649.3	27.4	14.8	6.4	8.4	12.6
Service-providing		86,229.9	2,826.6	1,428.4	840.6	587.8	1,398.3
Trade, transportation, and utilities[9]		25,564.9	1,155.5	659.5	380.7	278.8	496.0
Wholesale trade	42	5,724.8	246.6	146.8	80.2	66.6	99.9
Merchant wholesalers, durable goods	423	2,981.1	119.5	64.1	37.9	26.2	55.4
Merchant wholesalers, nondurable goods	424	2,012.4	110.0	72.6	36.6	36.0	37.4
Wholesale electronic markets and agents and brokers	425	731.3	17.1	10.0	5.7	4.3	7.1
Retail trade	44-45	15,209.1	603.1	314.2	175.9	138.4	288.9
Motor vehicle and parts dealers	441	1,911.9	87.1	38.6	25.8	12.8	48.6
Furniture and home furnishings stores	442	572.9	23.7	13.2	8.6	4.6	10.5
Electronics and appliance stores	443	531.0	10.8	5.8	2.9	2.9	5.0
Building material and garden equipment and supplies dealers	444	1,265.9	86.9	45.0	24.6	20.4	41.9
Food and beverage stores	445	2,809.0	125.7	67.8	40.8	27.1	57.9
Health and personal care stores	446	945.5	17.2	8.4	5.4	3.0	8.8
Gasoline stations	447	866.5	24.8	12.3	8.5	3.8	12.5
Clothing and clothing accessories stores	448	1,404.6	25.4	9.8	5.9	3.9	15.6
Sporting goods, hobby, book, and music stores	451	649.0	14.1	5.6	3.3	2.2	8.6
General merchandise stores	452	2,918.9	147.2	84.6	37.4	47.2	62.6
Miscellaneous store retailers	453	908.5	23.8	13.5	7.8	5.7	10.3
Nonstore retailers	454	425.5	16.5	9.7	4.9	4.8	6.8
Transportation and warehousing[9]	48-49	4,077.7	280.5	185.6	117.4	68.2	94.9

See footnotes at end of table.

Table 1-4. Numbers of nonfatal occupational injuries and illnesses by selected industries and case types, 2005. (*Continued*)

Industry[1]	NAICS code[2]	2005 Annual average employment[3]	Total recordable cases	Cases with days away from work, job transfer, or restriction: Total	Cases with days away from work[4]	Cases with job transfer or restriction	Other recordable cases
Air transportation	481	505.0	41.0	30.9	21.7	9.2	10.1
Rail transportation[9]	482	–	6.0	4.5	3.9	.6	1.5
Water transportation	483	58.0	2.4	1.5	1.1	.4	.9
Truck transportation	484	1,381.6	88.3	56.4	42.3	14.1	31.9
Transit and ground passenger transportation	485	384.3	18.1	10.2	6.8	3.4	7.9
Pipeline transportation	486	37.7	.8	.4	.2	.1	.5
Scenic and sightseeing transportation	487	27.6	1.0	.5	.4	.1	.6
Support activities for transportation	488	546.8	28.8	17.5	11.8	5.7	11.3
Couriers and messengers	492	559.1	49.1	34.0	17.0	17.0	15.1
Warehousing and storage	493	577.3	45.0	29.8	12.1	17.7	15.2
Utilities	22	553.3	25.3	12.9	7.2	5.7	12.4
Utilities	221	553.3	25.3	12.9	7.2	5.7	12.4
Information		3,064.0	58.9	30.9	20.7	10.2	27.9
Information	51	3,064.0	58.9	30.9	20.7	10.2	27.9
Publishing industries (except Internet)	511	904.9	17.5	8.9	5.7	3.1	8.6
Broadcasting (except Internet)	515	324.9	5.9	3.0	1.8	1.3	2.9
Internet publishing and broadcasting	516	30.7	.1	([10])	–	–	([7])
Telecommunications	517	998.7	25.2	14.4	10.2	4.2	10.9
Internet service providers, web search portals, and data processing services	518	380.3	4.1	1.9	1.1	.8	2.2
Other information services	519	49.6	.5	.2	.1	.1	.3
Financial activities		7,994.2	120.6	56.2	38.2	18.0	64.4
Finance and insurance	52	5,879.7	54.2	19.1	14.1	5.1	35.1
Monetary authorities - central bank	521	20.8	.4	.2	.1	.1	.2
Credit intermediation and related activities	522	2,852.1	27.3	9.5	7.2	2.2	17.8
Securities, commodity contracts, and other financial investments and related activities	523	786.7	2.3	1.1	.9	.1	1.2
Insurance carriers and related activities	524	2,132.4	23.3	8.1	5.6	2.5	15.2
Funds, trusts, and other financial vehicles	525	87.7	.9	.3	.2	.1	.6
Real estate and rental and leasing	53	2,114.5	66.4	37.1	24.2	12.9	29.3
Real estate	531	1,443.3	40.3	21.8	15.8	6.0	18.6
Rental and leasing services	532	645.1	25.8	15.2	8.2	7.0	10.6

See footnotes at end of table.

Table 1-4. Numbers of nonfatal occupational injuries and illnesses by selected industries and case types, 2005. (*Continued*)

Industry[1]	NAICS code[2]	2005 Annual average employment[3]	Total recordable cases	Cases with days away from work, job transfer, or restriction			Other recordable cases
				Total	Cases with days away from work[4]	Cases with job transfer or restriction	
Lessors of nonfinancial intangible assets (except copyrighted works)	533	26.1	0.2	0.1	0.1	(10)	0.1
Professional and business services		16,732.9	292.5	148.8	91.8	56.9	143.7
Professional, scientific, and technical services	54	6,979.3	87.7	38.4	24.8	13.6	49.3
Professional, scientific, and technical services	541	6,979.3	87.7	38.4	24.8	13.6	49.3
Management of companies and enterprises	55	1,730.9	39.3	20.8	9.7	11.1	18.5
Administrative and support and waste management and remediation services	56	8,022.7	165.4	89.5	57.3	32.2	75.9
Administrative and support services	561	7,688.3	141.1	73.3	48.0	25.3	67.8
Waste management and remediation services	562	334.4	24.3	16.2	9.3	6.9	8.1
Education and health services		16,385.7	705.9	333.2	186.4	146.8	372.7
Educational services	61	2,129.3	37.0	14.8	10.5	4.3	22.1
Educational services	611	2,129.3	37.0	14.8	10.5	4.3	22.1
Health care and social assistance	62	14,256.4	668.9	318.4	175.9	142.5	350.6
Ambulatory health care services	621	5,066.2	110.6	38.9	25.4	13.5	71.7
Hospitals	622	4,286.6	281.5	114.8	62.9	51.9	166.7
Nursing and residential care facilities	623	2,835.6	209.1	129.7	66.6	63.1	79.4
Social assistance	624	2,068.1	67.7	35.0	21.0	14.0	32.7
Leisure and hospitality		12,687.5	398.5	154.9	93.9	61.0	243.6
Arts, entertainment, and recreation	71	1,865.5	72.1	34.1	18.2	15.9	38.0
Performing arts, spectator sports, and related industries	711	379.3	19.5	6.7	4.2	–	–
Museums, historical sites, and similar institutions	712	118.4	4.3	1.7	1.0	.7	2.7
Amusement, gambling, and recreation industries	713	1,367.8	48.2	25.8	13.1	12.7	22.4
Accommodation and food services	72	10,822.0	326.4	120.8	75.7	45.1	205.6
Accommodation	721	1,806.8	85.1	46.4	23.5	22.9	38.7
Food services and drinking places	722	9,015.3	241.3	74.3	52.1	22.2	167.0
Other services		3,800.7	94.8	44.8	28.8	16.0	49.9

See footnotes at end of table.

Table 1-4. Numbers of nonfatal occupational injuries and illnesses by selected industries and case types, 2005. (*Continued*)

Industry[1]	NAICS code[2]	2005 Annual average employment[3]	Total recordable cases	Cases with days away from work, job transfer, or restriction			Other recordable cases
				Total	Cases with days away from work[4]	Cases with job transfer or restriction	
Other services, except public administration	81	3,800.7	94.8	44.8	28.8	16.0	49.9
Repair and maintenance	811	1,231.8	44.3	20.6	14.3	6.3	23.8
Personal and laundry services	812	1,271.6	26.1	14.9	7.9	7.0	11.2
Religious, grantmaking, civic, professional, and similar organizations	813	1,297.3	24.3	9.4	6.6	2.8	14.9

1 Totals include data for industries not shown separately.
2 *North American Industry Classification System* — United States, 2002
3 Employment is expressed as an annual average and is derived primarily from the BLS-Quarterly Census of Employment and Wages (QCEW) program.
4 Days-away-from-work cases include those that result in days away from work with or without job transfer or restriction.
5 Excludes farms with fewer than 11 employees.
6 Data for Mining (Sector 21 in the *North American Industry Classification System*-- United States, 2002) include establishments not governed by the Mine Safety and Health Administration rules and reporting, such as those in Oil and Gas Extraction and related support activities. Data for mining operators in coal, metal, and nonmetal mining are provided to BLS by the Mine Safety and Health Administration, U.S. Department of Labor. Independent mining contractors are excluded from the coal, metal, and nonmetal mining industries. These data do not reflect the changes the Occupational Safety and Health Administration made to its recordkeeping requirements effective January 1, 2002; therefore, estimates for these industries are not comparable to estimates in other industries.
7 Fewer than 50 cases.
8 Data for mining operators in this industry are provided to BLS by the Mine Safety and Health Administration, U.S. Department of Labor. Independent mining contractors are excluded. These data do not reflect the changes the Occupational Safety and Health Administration made to its recordkeeping requirements effective January 1, 2002; therefore, estimates for these industries are not comparable to estimates in other industries.
9 Data for employers in railroad transportation are provided to BLS by the Federal Railroad Administration, U.S. Department of Transportation.
10 Fewer than 15 cases.

NOTE: Because of rounding, components may not add to totals. Dash indicates data not available.
SOURCE: Bureau of Labor Statistics, U.S. Department of Labor

Figure 1-2. Incidence rates for cases with days away from work, job transfer, or restriction by case type and selected industry sector, 2005.

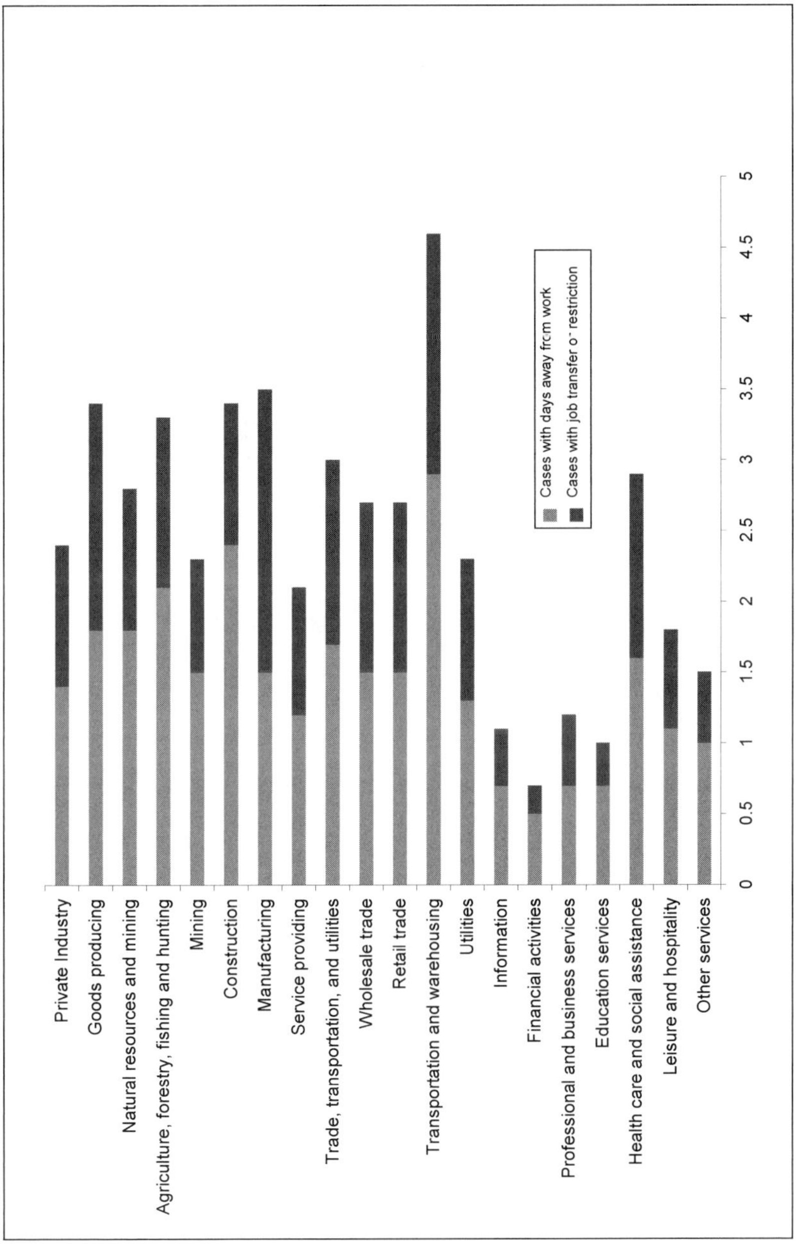

Source: BLS

Table 1-5. Incidence and number of nonfatal occupational injuries by selected industries, 2005.

Industry[2]	NAICS code[3]	2005 Annual average employment[4] (thousands)	Incidence rate	Number of cases (thousands)
Private industry[5]		109,127.0	4.4	3,971.7
Goods-producing[5]		22,897.1	5.7	1,279.5
Natural resources and mining[5,6]		1,517.7	4.8	73.0
Agriculture, forestry, fishing and hunting[5]	11	969.5	5.7	51.4
Crop production[5]	111	428.3	5.3	21.7
Animal production[5]	112	144.7	7.9	12.7
Forestry and logging	113	71.6	5.9	3.8
Fishing, hunting and trapping	114	9.2	3.3	.3
Support activities for agriculture and forestry	115	315.7	5.1	13.0
Mining[6]	21	548.2	3.5	21.6
Oil and gas extraction	211	124.4	2.0	2.5
Mining (except oil and gas)[7]	212	209.2	4.0	9.8
Support activities for mining	213	214.6	3.8	9.3
Construction		7,166.6	6.2	404.6
Construction	23	7,166.6	6.2	404.6
Construction of buildings	236	1,682.8	5.2	79.7
Heavy and civil engineering construction	237	921.5	5.3	48.6
Specialty trade contractors	238	4,562.3	6.7	276.3
Manufacturing		14,212.8	5.6	801.9
Manufacturing	31-33	14,212.8	5.6	801.9
Food manufacturing	311	1,479.2	6.5	96.0
Beverage and tobacco product manufacturing	312	192.0	7.8	14.9
Textile mills	313	222.4	3.8	8.4
Textile product mills	314	171.2	4.6	7.5
Apparel manufacturing	315	263.7	2.9	7.0
Leather and allied product manufacturing	316	40.0	5.3	2.0
Wood product manufacturing	321	556.8	8.7	48.8
Paper manufacturing	322	486.2	4.0	20.4
Printing and related support activities	323	648.3	3.9	24.1
Petroleum and coal products manufacturing	324	112.3	2.8	3.3
Chemical manufacturing	325	874.8	2.8	24.4
Plastics and rubber products manufacturing	326	800.8	6.7	54.6
Nonmetallic mineral product manufacturing	327	503.0	7.5	39.1

See footnotes at end of table.

Table 1-5. Incidence and number of nonfatal occupational injuries by selected industries, 2005. (*Continued*)

Industry[2]	NAICS code[3]	2005 Annual average employment[4] (thousands)	Incidence rate	Number of cases (thousands)
Primary metal manufacturing	331	466.1	8.3	40.7
Fabricated metal product manufacturing	332	1,509.5	7.5	113.7
Machinery manufacturing	333	1,153.7	6.1	70.9
Computer and electronic product manufacturing	334	1,310.7	1.7	22.4
Electrical equipment, appliance, and component manufacturing	335	436.9	4.8	20.8
Transportation equipment manufacturing	336	1,770.1	6.8	120.7
Furniture and related product manufacturing	337	565.7	6.7	37.4
Miscellaneous manufacturing	339	649.3	4.0	24.8
Service-providing		86,229.9	3.9	2,692.2
Trade, transportation, and utilities[8]		25,564.9	5.1	1,119.8
Wholesale trade	42	5,724.8	4.4	239.8
Merchant wholesalers, durable goods	423	2,981.1	4.0	116.1
Merchant wholesalers, nondurable goods	424	2,012.4	5.6	107.1
Wholesale electronic markets and agents and brokers	425	731.3	2.5	16.6
Retail trade	44-45	15,209.1	4.9	587.6
Motor vehicle and parts dealers	441	1,911.9	4.7	85.0
Furniture and home furnishings stores	442	572.9	4.9	23.4
Electronics and appliance stores	443	531.0	2.4	10.7
Building material and garden equipment and supplies dealers	444	1,265.9	7.4	86.1
Food and beverage stores	445	2,809.0	5.8	122.1
Health and personal care stores	446	945.5	2.3	16.6
Gasoline stations	447	866.5	3.4	24.2
Clothing and clothing accessories stores	448	1,404.6	2.6	25.0
Sporting goods, hobby, book, and music stores	451	649.0	3.2	13.9
General merchandise stores	452	2,918.9	6.5	142.0
Miscellaneous store retailers	453	908.5	3.5	23.2
Nonstore retailers	454	425.5	4.3	15.4
Transportation and warehousing[8]	48-49	4,077.7	6.7	269.6
Air transportation	481	505.0	9.4	38.7
Rail transportation[8]	482	–	2.4	5.7
Water transportation	483	58.0	3.6	2.2
Truck transportation	484	1,381.6	6.0	87.0
Transit and ground passenger transportation	485	384.3	6.0	17.5
Pipeline transportation	486	37.7	1.7	.6
Scenic and sightseeing transportation	487	27.6	5.2	1.0
Support activities for transportation	488	546.8	5.3	27.7
Couriers and messengers	492	559.1	10.8	45.6

See footnotes at end of table.

Table 1-5. Incidence and number of nonfatal occupational injuries by selected industries, 2005. (*Continued*)

Industry[2]	NAICS code[3]	2005 Annual average employment[4] (thousands)	Incidence rate	Number of cases (thousands)
Warehousing and storage	493	577.3	7.9	43.5
Utilities	22	553.3	4.2	22.8
Utilities	221	553.3	4.2	22.8
Information		3,064.0	1.9	53.6
Information	51	3,064.0	1.9	53.6
Publishing industries (except Internet)	511	904.9	2.0	16.4
Broadcasting (except Internet)	515	324.9	1.9	5.6
Internet publishing and broadcasting	516	30.7	.1	([9])
Telecommunications	517	998.7	2.3	22.3
Internet service providers, web search portals, and data processing services	518	380.3	1.0	3.7
Other information services	519	49.6	1.1	.4
Financial activities		7,994.2	1.5	111.1
Finance and insurance	52	5,879.7	.9	47.1
Monetary authorities - central bank	521	20.8	1.8	.3
Credit intermediation and related activities	522	2,852.1	.9	24.6
Securities, commodity contracts, and other financial investments and related activities	523	786.7	.3	2.1
Insurance carriers and related activities	524	2,132.4	1.0	19.1
Funds, trusts, and other financial vehicles	525	87.7	1.1	.9
Real estate and rental and leasing	53	2,114.5	3.6	64.1
Real estate	531	1,443.3	3.2	38.7
Rental and leasing services	532	645.1	4.5	25.2
Lessors of nonfinancial intangible assets (except copyrighted works)	533	26.1	.7	.2
Professional and business services		16,732.9	2.2	275.3
Professional, scientific, and technical services	54	6,979.3	1.3	83.0
Professional, scientific, and technical services	541	6,979.3	1.3	83.0
Management of companies and enterprises	55	1,730.9	2.3	36.9
Administrative and support and waste management and remediation services	56	8,022.7	3.5	155.5
Administrative and support services	561	7,688.3	3.2	131.8
Waste management and remediation services	562	334.4	6.9	23.7

See footnotes at end of table.

Table 1-5. Incidence and number of nonfatal occupational injuries by selected industries, 2005. (*Continued*)

Industry[2]	NAICS code[3]	2005 Annual average employment[4] (thousands)	Incidence rate	Number of cases (thousands)
Education and health services		16,385.7	5.1	658.9
Educational services	61	2,129.3	2.3	35.0
Educational services	611	2,129.3	2.3	35.0
Health care and social assistance	62	14,256.4	5.5	623.9
Ambulatory health care services	621	5,066.2	2.5	101.0
Hospitals	622	4,286.6	7.5	258.7
Nursing and residential care facilities	623	2,835.6	8.7	199.9
Social assistance	624	2,068.1	4.1	64.4
Leisure and hospitality		12,687.5	4.5	383.9
Arts, entertainment, and recreation	71	1,865.5	5.8	67.8
Performing arts, spectator sports, and related industries	711	379.3	8.1	19.1
Museums, historical sites, and similar institutions	712	118.4	4.9	4.0
Amusement, gambling, and recreation industries	713	1,367.8	5.2	44.7
Accommodation and food services	72	10,822.0	4.3	316.2
Accommodation	721	1,806.8	5.7	80.2
Food services and drinking places	722	9,015.3	4.0	236.0
Other services		3,800.7	3.0	89.6
Other services, except public administration	81	3,800.7	3.0	89.6

See footnotes at end of table.

Table 1-5. Incidence and number of nonfatal occupational injuries by selected industries, 2005. (*Continued*)

Industry[2]	NAICS code[3]	2005 Annual average employment[4] (thousands)	Incidence rate	Number of cases (thousands)
Repair and maintenance	811	1,231.8	3.9	43.2
Personal and laundry services	812	1,271.6	2.6	24.9
Religious, grantmaking, civic, professional, and similar organizations	813	1,297.3	2.4	21.5

[1] The incidence rates represent the number of injuries per 100 full-time workers and were calculated as: (N/EH) x 200,000, where

N = number of injuries
EH = total hours worked by all employees during the calendar year
200,000 = base for 100 equivalent full-time workers (working 40 hours per week, 50 weeks per year)

[2] Totals include data for industries not shown separately.
[3] *North American Industry Classification System* — United States, 2002
[4] Employment is expressed as an annual average and is derived primarily from the BLS-Quarterly Census of Employment and Wages (QCEW) program.
[5] Excludes farms with fewer than 11 employees.
[6] Data for Mining (Sector 21 in the *North American Industry Classification System*--United States, 2002) include establishments not governed by the Mine Safety and Health Administration rules and reporting, such as those in Oil and Gas Extraction and related support activities. Data for mining operators in coal, metal, and nonmetal mining are provided to BLS by the Mine Safety and Health Administration, U.S. Department of Labor. Independent mining contractors are excluded from the coal, metal, and nonmetal mining industries. These data do not reflect the changes the Occupational Safety and Health Administration made to its recordkeeping requirements effective January 1, 2002; therefore, estimates for these industries are not comparable to estimates in other industries.
[7] Data for mining operators in this industry are provided to BLS by the Mine Safety and Health Administration, U.S. Department of Labor. Independent mining contractors are excluded. These data do not reflect the changes the Occupational Safety and Health Administration made to its recordkeeping requirements effective January 1, 2002; therefore, estimates for these industries are not comparable to estimates in other industries.
[8] Data for employers in railroad transportation are provided to BLS by the Federal Railroad Administration, U.S. Department of Transportation.
[9] Fewer than 50 cases.

NOTE: Because of rounding, components may not add to totals. Dash indicates data not available.
SOURCE: Bureau of Labor Statistics, U.S. Department of Labor

- **Illnesses**—Workplace illnesses accounted for fewer than 6 percent of the 4.1 million injury and illness cases in 2005, unchanged from 2004. There were 242,500 newly reported cases of occupational illnesses in private industry in 2005, relatively unchanged from the 249,000 cases in 2004. Service-providing industries accounted for approximately 55 percent of these cases, while goods-producing industries accounted for 45 percent. The manufacturing sector accounted for nearly 39 percent of all newly reported cases of occupational illnesses (see Table 1-6 and Figure 1-3). The category, "All other illnesses," accounted for 63 percent of total illness cases in 2005, compared to over 65 percent in 2004. Private industry experienced a statistically significant decline in 2005 in both the number of cases and the incidence rate of "All other illnesses"; the remaining categories of illness, however, remained relatively unchanged. Beginning with the 2004 calendar year, the Occupational Safety and Health Administration (OSHA) included "Hearing loss" as a separate illness category. Hearing loss accounted for 11 percent of all illnesses in 2005, relatively unchanged from 2004. Prior to 2004, hearing loss cases were included in the "All other illnesses" category.

- The survey measures the number of new work-related illnesses cases that are recognized, diagnosed, and reported during the year. The bulk of the reported new illnesses come directly from workplace activity such as contact dermatitis or carpal tunnel syndrome. Long-term latent illnesses such as illnesses caused by exposure to carcinogens are difficult to relate to the workplace and are not adequately recognized and reported.

Industry Sectors Reporting

- Agriculture, forestry, fishing and hunting—This industry sector accounted for slightly less than 1 percent of private sector employment and slightly more than 1 percent of all injury/illness cases in 2005. While the incidence rate for this industry sector was significantly higher than that of private industry in 2005, the rate, as well as the number of injury/illness cases, remained relatively unchanged from 2004.

Table 1-6. Incidence rates[1] and numbers of nonfatal occupational illnesses by major industry sector and category of illness, 2005.

Industry sector	Total cases	Skin diseases or disorders	Respiratory conditions	Poisonings	Hearing loss	All other illnesses
	Incidence rates per 10,000 full-time workers					
Private industry[2]	26.7	4.4	2.2	0.3	3.0	16.8
Goods-producing[2]	48.5	6.3	2.6	.3	10.3	29.1
Natural resources and mining[2,3]	25.5	8.3	2.8	.6	1.9	11.8
Construction	15.7	4.3	1.7	.3	.5	8.9
Manufacturing	66.1	7.0	–	.3	15.7	40.2
Service-providing	19.6	3.8	2.1	.3	.6	12.8
Trade, transportation, and utilities[4]	16.2	2.6	1.4	.3	1.4	10.5
Information	19.1	2.1	1.0	.2	.9	14.8
Financial activities	13.1	1.1	1.5	.1	(5)	10.4
Professional and business services	13.8	3.3	1.3	.3	.4	8.5
Education and health services	36.7	6.6	4.8	.2	.1	25.0
Leisure and hospitality	17.2	5.8	2.4	.4	.1	8.6
Other services, except public administration	17.4	5.1	1.1	.6	.2	10.3
	Numbers of illnesses in thousands					
Private industry[2]	242.5	40.1	20.2	2.8	26.9	152.4
Goods-producing[2]	108.1	13.9	5.7	.8	22.9	64.8
Natural resources and mining[2,3]	3.9	1.3	.4	.1	.3	1.8
Construction	10.3	2.8	1.1	.2	.3	5.9
Manufacturing	94.0	9.9	–	.5	22.3	57.2
Service-providing	134.4	26.2	14.5	2.0	4.1	87.6
Trade, transportation, and utilities[4]	35.8	5.8	3.1	.8	3.0	23.1
Information	5.3	.6	.3	.1	.3	4.1
Financial activities	9.5	.8	1.1	.1	(6)	7.5
Professional and business services	17.1	4.1	1.6	.3	.5	10.5
Education and health services	47.0	8.4	6.1	.3	.1	32.0
Leisure and hospitality	14.6	4.9	2.0	.3	.1	7.3
Other services, except public administration	5.1	1.5	.3	.2	.1	3.0

[1] The incidence rates represent the number of illnesses per 10,000 full-time workers and were calculated as: (N/EH) x 20,000,000, where

N = number of illnesses
EH = total hours worked by all employees during the calendar year
20,000,000 = base for 10,000 equivalent full-time workers (working 40 hours per week, 50 weeks per year)

[2] Excludes farms with fewer than 11 employees.

[3] Data for Mining (Sector 21 in the *North American Industry Classification System*-- United States, 2002) include establishments not governed by the Mine Safety and Health Administration rules and reporting, such as those in Oil and Gas Extraction and related support activities. Data for mining operators in coal, metal, and nonmetal mining are provided to BLS by the Mine Safety and Health Administration, U.S. Department of Labor. Independent mining contractors are excluded from the coal, metal, and nonmetal mining industries. These data do not reflect the changes the Occupational Safety and Health Administration made to its recordkeeping requirements effective January 1, 2002; therefore, estimates for these industries are not comparable to estimates in other industries.

[4] Data for employers in railroad transportation are provided to BLS by the Federal Railroad Administration, U.S. Department of Transportation.

[5] Incidence rate less than 0.05.

[6] Fewer than 50 cases.

NOTE: Because of rounding, components may not add to totals.
SOURCE: Bureau of Labor Statistics, U.S. Department of Labor

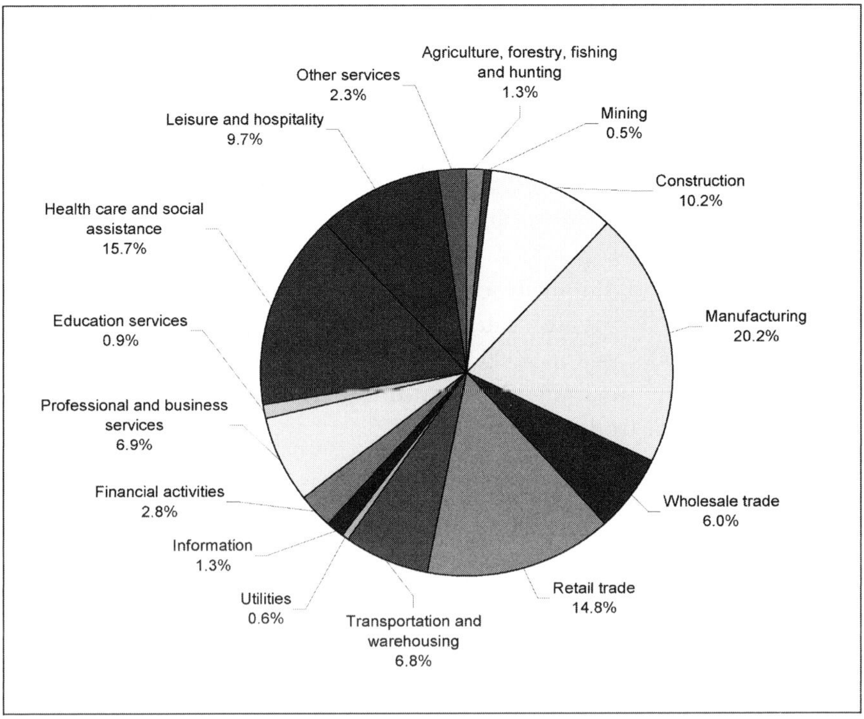

Figure 1-3. Percent of nonfatal workplace illnesses by industry sector, 2005. Source: BLS

- Mining—The incidence rate of injury/illnesses in this industry sector was significantly lower than that of private industry in 2005. The number of injury/illness cases, as well as the incidence rate in this sector, remained unchanged compared to 2004. However, the incidence rate for cases that involved days of job transfer or restriction rose significantly in 2005 compared to 2004.

- Construction—The incidence rate of injury/illnesses for this industry sector was unchanged from 2004. The specialty trade construction sector (masonry, roofing, electrical, plumbing and painting) reported the highest rate of injury/illnesses. The number of cases and the incidence rates remained relatively unchanged in 2005 for actual building construction and heavy and civil engineering construction.

- Manufacturing—Although this industry sector accounted for roughly 13 percent of private sector employment, more than 1 in 5 injury/illness cases reported in private industry, and nearly 2 in 5 illnesses occurred in manufacturing in 2005. While the incidence rate of injury/illnesses for this industry sector declined in 2005, the rate remained significantly higher than that of overall private industry for that reporting year.

- Wholesale and Retail Trade—This sector accounted for approximately 1 in 5 private sector employees and an equal number of injury/illness cases in 2005. While the incidence rate for wholesale trade remain unchanged, the number and rate of injury/illness cases in retail trade declined in 2005.

- Transportation and Warehousing—This sector includes passenger and freight transport via air, water, rail and ground, as well as warehousing and storage of goods, sightseeing transportation and various related support activities. While the incidence rates of injury/illness in this sector remained relatively unchanged compared to 2004, the rate for this sector was significantly higher than that of total private industry in 2005.

- Utilities—This relatively small industry sector (electric power, natural gas, water and sewage removal) accounted for only one-half of one percent of private industry employment and injury/illness cases in 2005. The injury/illness incidence rate for utilities was not significantly different from 2004; however, the number of cases reported in the industry did decline significantly in 2005.

- Information—Data for this sector, which includes industries such as telecommunications, Internet service providers, publishing and broadcasting, and motion picture and sound recording, remained relatively unchanged from 2004 but significantly lower than the rate for private industry in general.

- Financial Activities—This sector encompasses finance and insurance and real estate and rental/leasing. The majority of the injuries/illnesses in financial activities occurred in real estate and rental and leasing. The real estate and rental/leasing experience a significantly higher rate of total recordable injuries and illnesses than did the finance and insurance sectors.

- Professional and Business Services—The three sectors in this industry include professional, scientific and technical services; management of companies and enterprises; and administrative and support and waste management and remediation services. The injuries and illness rate was well below that of all private industry and remained unchanged from 2004.

- Education Services and Health Care/Social Assistance—The incidence rate for education services remained relatively unchanged in 2005, while the rate in health care declined. Health care and social assistance in comprised of ambulatory health care services, hospitals, nursing and residential care facilities, and social assistance. However, hospitals topped the list of all industries reporting 100,000 or more cases. The rate of illnesses experienced by workers in hospitals was 66.2 per 10,000 full time workers versus 26.7 cases for all private industry in 2005.

- Leisure and Hospitality—This industry includes arts, entertainment, recreation, accommodation and food services. While the injury/illness rates for these sectors remained relatively unchanged from 2004, the rate for arts, entertainment and recreation was higher than the rate for accommodation and food services in 2005.

- Others—This diverse industry sector includes everything from auto repair and maintenance to cemeteries and crematories, nail and beauty salons, dry cleaning, and religious, civic, professional, and similar organizations. The incidence rate of injury/illnesses for this industry sector remained unchanged in 2005. However, repair and maintenance accounted for nearly half of all injury/illness cases reported in this sector.

Implications for Employers

What do these statistics mean for employers?

The strongest lesson of these findings is that safety is a primary workforce issue. Whether viewed alone, or in tandem with the indirect costs of replacing injured workers with temporary staff, overtime pay,

Table 1-7. Incidence rates[1] of nonfatal occupational injuries and illnesses by major industry sector and selected case types, 2005.

Industry sector	Total recordable cases			Cases with days away from work, job transfer, or restriction: Total			Cases with days away from work[2]			Cases with job transfer or restriction			Other recordable cases		
	2003	2004	2005	2003	2004	2005	2003	2004	2005	2003	2004	2005	2003	2004	2005
Private industry[3]	5.0	4.8	4.6	2.6	2.5	2.4	1.5	1.4	1.4	1.1	1.1	1.0	2.4	2.3	2.2
Goods-producing[3]	6.7	6.5	6.2	3.7	3.5	3.4	1.9	1.9	1.8	1.8	1.7	1.6	3.0	2.9	2.8
Natural resources and mining[3,4]	5.1	5.3	5.1	2.8	3.1	2.9	1.8	2.0	1.8	1.0	1.1	1.0	2.3	2.2	2.2
Construction	6.8	6.4	6.3	3.6	3.4	3.4	2.6	2.4	2.4	1.0	.9	1.0	3.2	3.0	2.9
Manufacturing	6.8	6.6	6.3	3.8	3.6	3.5	1.6	1.6	1.5	2.2	2.1	2.0	3.1	3.0	2.8
Service-providing	4.4	4.2	4.1	2.3	2.2	2.1	1.4	1.3	1.2	.9	.9	.9	2.1	2.1	2.0
Trade, transportation, and utilities[5]	5.5	5.5	5.2	3.2	3.1	3.0	1.9	1.8	1.7	1.3	1.3	1.3	2.4	2.3	2.2
Information	2.2	2.0	2.1	1.1	1.1	1.1	.8	.8	.7	.3	.4	.4	1.1	.9	1.0
Financial activities	1.7	1.6	1.7	.8	.7	.8	.6	.5	.5	.2	.2	.2	.9	.9	.9
Professional and business services	2.5	2.4	2.4	1.4	1.3	1.2	.9	.8	.7	.5	.5	.5	1.1	1.1	1.2
Education and health services	6.0	5.8	5.5	2.9	2.7	2.6	1.6	1.5	1.5	1.3	1.2	1.1	3.1	3.1	2.9
Leisure and hospitality	5.1	4.7	4.7	2.1	1.9	1.8	1.3	1.1	1.1	.8	.8	.7	3.0	2.8	2.9
Other services, except public administration	3.4	3.2	3.2	1.7	1.6	1.5	1.1	1.1	1.0	.6	.5	.5	1.7	1.6	1.7

[1] The incidence rates represent the number of injuries and illnesses per 100 full-time workers and were calculated as: (N/EH) x 200,000, where

N = number of injuries and illnesses
EH = total hours worked by all employees during the calendar year
200,000 = base for 100 equivalent full-time workers (working 40 hours per week, 50 weeks per year)

[2] Days-away-from-work cases include those that result in days away from work with or without job transfer or restriction.
[3] Excludes farms with fewer than 11 employees.
[4] Data for Mining (Sector 21 in the *North American Industry Classification System*-- United States, 2002) include establishments not governed by the Mine Safety and Health Administration rules and reporting, such as those in Oil and Gas Extraction and related support activities. Data for mining operators in coal, metal, and nonmetal mining are provided to BLS by the Mine Safety and Health Administration, U.S. Department of Labor. Independent mining contractors are excluded from the coal, metal, and nonmetal mining industries. These data do not reflect the changes the Occupational Safety and Health Administration made to its recordkeeping requirements effective January 1, 2002; therefore, estimates for these industries are not comparable to estimates in other industries.
[5] Data for employers in railroad transportation are provided to BLS by the Federal Railroad Administration, U.S. Department of Transportation.

NOTE: Because of rounding, components may not add to totals.
SOURCE: Bureau of Labor Statistics, U.S. Department of Labor

downtime and lost productivity, the consequences of poor safety can be a crippling debit to a company's bottom line.

The challenge of managing safety in the workplace is rooted in integrating the safety function into the full scope of company operations, including strategic and tactical planning, financial forecasting, budgeting and cost containment.

As part of the scope of a company's operations, worker safety becomes as much a risk management issue as any other facet of the company's operations.

RISK MANAGEMENT: THE KEY TO COST CONTAINMENT

Today's employers must cope with the acceleration of safety and health regulations in the workplace; proposed legislation to extend coverage to industries and employers previously exempted; increased employer penalties for non-compliance; revisions to longstanding guidelines; and increased employee litigation.

Workplace accidents can happen. Illnesses will occur.

Such incidents, however, can be prevented or minimized through effective risk management programming.

Risk management involves:

1) Identifying hazards and potential hazards.
2) Taking action to reduce and/or eliminated them.

Taking such action minimized employers exposure to liability, losses resulting from workers' compensation claims, and hidden losses resulting from production downtime, employee absence and replacement and/or job reassignment.

Cost Containment Strategies

All workers' compensation losses result from workplace accident or illness. Controlling these losses involve undertaking various strategies that include:

- Developing management awareness programs.
- Installing employee involvement programming.
- Investigating accidents and incidents.
- Conducting safety audits.

- Managing the claims process.
- Managing litigation.

Develop Management Awareness Programs

Facility managers and other people responsible for their companies; risk management efforts should help the management team gain a clear understanding of the net effects of workers' compensation costs.

The "real" costs of a seemingly "minor" claim, for example, include more than the total dollar figure listed on the company's loss report. The "hidden" costs of production downtime, employee recruitment, replacement and training expenses incurred, must be factored in to determine actual per-claim costs.

Quantifying net costs of workers' compensation claims helps all members of the management team realize the importance of committing to cost containment efforts. Disseminating cost breakdowns at management and division meetings, etc., "drives the message home" even further.

Install Employee Involvement Programming

For cost containment efforts to work, employees at all levels must be included.

A company safety committee, which is comprised of both managers and employees, provides an effective vehicle for employee involvement. Because the employees are closest to work situations, they can speak directly to safety conditions. In the Committee, they have a vehicle to bring their concerns directly to management. In turn, management is provided the opportunity to address employee concerns and promote employee well-being.

A second benefit of employee involvement is the increased level of commitment to safe work practices. Peer support for program goals and objectives that are generated, at least in part by employees, is easier to obtain.

Investigate Accidents And Incidents

A thorough investigation should be conduced promptly following any accident or injury. By determining the causes that led up to the accident, employers will have much of the information that is needed to prevent a recurrence of the accident or injury.

Prompt follow-up investigations demonstrate to all employees that management is serious about workplace safety.

"Near miss" incidents should be promptly and thoroughly investigated as well. The only thing that separates a near-miss incident from an actual incident or injury is luck. And luck runs out sooner or later.

While it is essential that all investigations are promptly conducted following an accident, an effective risk management program requires taking preventative measures. Anticipating potential problems can reduce or eliminate conditions that can cause accidents.

Conduct Safety Audits

Conducting a facility-wide audit is an integral component of the cost containment initiative. It is also an essential component of overall safety and health programming. In fact, OSHA has identified worksite analysis (e.g., audit) as one of the four basic elements of effective safety and health programming.

Whether companies choose to self-audit or secure an independent analysis is not as important as actually undergoing the audit itself. While independent and in-house audits have their own distinct advantages, both should minimal include an evaluation and/or review of the company's:

- Safety and health policy
- Safety committee activities.
- Safety and health training programs (all employee levels).
- Accident and incident investigation procedures and practices.
- Record-keeping procedures and practices.
- Compliance requirements (all applicable Federal and State requirements).

Manage Claims

Managing the claims administration process involves periodic file/case review. A representative number of open cases randomly selected closed cases, and all high-threshold claims should be reviewed. In doing so, specific attention should focus on the indicators that suggest problems are occurring in the claims administration process. For example, any increase in the slow-down rate of case closings may indicate problems. Other "red-flag" indicators include increases in the number of litigated claims, newly reported cases, reopened claims, late-reported claims and missed reimbursements.

While this list of indicators is not all inclusive, it provides a starting point for effectively managing one of the most critical elements of a cost containment program.

Manage Litigation

Generally speaking, few workers' compensation claims reach the hearing stage. When they do, however, they can be costly. In addition to settlement costs and attorneys' fees, litigation expenses must be factored in to arrive at actual per-litigated-case costs.

In those cases which merit settlement, companies may want to use appropriately trained and experienced in-house personnel and/or adjusters to negotiate any warranted settlement.

Sources

Gustin, Joseph F., *Safety Management: A Guide For Facility Managers*, New York: UpWord Publishing, Inc., 1996.

Gustin, Joseph F., *Disaster and Recovery Planning: A Guide For Facility Managers, 4th ed.*, Lilburn, GA: The Fairmont Press, Inc., 2007.

The OSH Act of 1970 (December 29, 1970), PL 91-596, 29 USC 651

United States Department of Labor, Bureau of Labor Statistics. *Workplace Injuries And Illnesses in 2005*, USDL 06-1816, October, 2006

United States Department of Labor, Bureau of Labor Statistics. *National Census Of Fatal Occupational Injuries in 2005*, USDL 06-1364, August, 2006.

United States Department of Labor, Bureau of Labor Statistics. *Nonfatal Occupation Injuries And Illnesses Requiring Days Away From Work, 2005*, USDL 06-1982, November, 2006.

United States Department of Labor, Occupational Safety and Health Administration, *All About OSHA*, OSHA 3302-06N, 2006.

Chapter 2
Rights and Responsibilities

The Occupational Safety and Health (OSH) Act of 1970 created the Occupational Safety and Health Administration, (OSHA), within the Department of Labor (DOL). The Act encouraged employers and employees to reduce workplace hazards and to implement safety and health programs. The Act also gave employees many new rights and responsibilities. This chapter discusses the rights and responsibilities of both employees and employers.

RIGHTS, RESPONSIBILITIES AND EXPECTATIONS

Employers, employees and regulatory agencies all have a vested interest in providing and maintaining a safe work environment. However, each focuses on various aspects of safety. Each has a separate list of rights, responsibilities and expectations as well. It may seem that employees maintain a disproportionate share of rights, but because of the strong vested common interest in establishing a work environment free of safety hazards, this should not detract from the opportunity for employer and employee to work together to plan and install an aggressive safety and healthy program. In fact, the effective safety and health program will take into account the expectations, rights and responsibilities of all parties involved.

OSHA Inspection Activity Remains Vigorous
OSHA continues to maintain its high level of annual inspection activity. In 2005, OSHA conducted 38,714 total inspections, exceeding its goal of 37,700. Unprogrammed, or unscheduled, inspections showed an increase over the previous year. OSHA responded to more employee complaints and conducted more inspections based on referrals from other agencies than the previous year. Programmed inspections continue to focus on high hazard industries, which have the highest lost-time injury rates. Note: Programmed inspections are worksite inspections which are scheduled based upon objective or neutral criteria. These inspections do not include imminent danger, fatality/catastrophe, and formal complaints

OSHA Inspection Statistics	*FY2001*	*FY2002*	*FY2003*	*FY2004*	*FY2005*	*% Change 2001-2005*
Total Inspections	35,974	37,614	39,817	39,167	38,714	+7.6%
Total Programmed Inspections	17,946	20,539	22,436	21,576	21,404	+19.3%
Total Unprogrammed (Unscheduled) Inspections	18,027	17,075	17,381	17,590	17,310	–4.0%
Fatality Investigations	1,130	1,134	1,021	1,060	1,114	–1.4%
Complaints	8,374	7,896	7,969	8,062	7,716	–7.9%
Referrals	4,434	4,447	4,472	4,585	4,787	+8.0%
Other	4,089	3,598	3,880	3,829	4,807	+17.6

Employer Rights

While the issue of employer rights v. employer responsibilities remains controversial, there are several key rights that employers are granted. These rights include:

- The right to comment on proposed regulations and amendments to existing regulations;
- The right to contest citations, proposed abatement periods and/or penalties; and
- The right of variance.

The Right to Comment

Prior to issuing, amending or deleting any regulation, OSHA publishes notice in the Federal Register, so that any interested party, including employers, may comment on any or all aspects of the regulation.

The Right to Contest

When workplace hazards are discovered by OSHA officers during an inspection, employers may be issued citations that list the alleged violation(s). Penalties, as well as any abatement periods, may also be proposed. Employers may contest the alleged violations, abatement period and any proposed penalty.

The Right of Variance

There are times when employers cannot fully comply with a new safety or health standard within the prescribed time frame. There are also situations where employers believe that their methods, equipment or facilities are equal to, or exceed, OSHA's requirements. In these instances, employers may apply for either a temporary or permanent variance.

Employers may also apply for an experimental or defense variance, and multi-state employers may apply for a multi-state variance.

Types of Variances

Temporary Variance

Employers may apply for a temporary variance from an OSHA standard's prescribed compliance date, when there is a shortage of necessary personnel, materials or equipment; or when the necessary construction or alteration to a facility cannot be completed in time.

To be eligible for a temporary variance, the employer must implement an effective program for meeting compliance to the standard as quickly as possible. The employer must also demonstrate that steps will be taken to protect employees, including:

- Identifying the standard, or regulation, or that portion of the standard or regulation from which variance is requested.
- Stating the reason(s) why the employer cannot comply by the effective date of the standard or regulation.
- Providing documentation of the measures the employer has already taken, as well as those measures that will be taken (including dates), to comply with the standard or regulation.
- Certifying that employees have been informed of the application for variance and that a copy has been given to the employees' authorized representative.
- Posting a summary of the application for variance wherever notices are normally posted in the workplace.

- Informing employees of their right to request a hearing on the variance application.

Temporary variances to any OSHA standard or regulation are not granted to any employer who maintains that the cost for compliance is prohibitive.

Employers who seek a temporary variance may also apply to OSHA for an interim order. The purpose of an interim order is to enable the employer to continue operating under existing conditions until a decision is made on the variance application.

Permanent Variances

Employers may apply for a permanent variance to standards or regulations if they can prove that their methods, conditions, operations, practices or processes can meet or exceed the standard's requirements. Contingent upon the submitted evidence and a work site inspection to confirm the merit of the request, a permanent variance may be granted.

In those situations where a permanent variance is granted, the final variance order will detail the employer's specific responsibilities and requirements, and explain how the employer's methods vary from the OSHA standard or regulation.

In applying for a permanent variance, employers must notify employees of the application and inform them of their right to request a hearing. All applications for permanent variances must contain essentially the same information required for temporary variances.

Experimental Variance

Experimental variances are granted to employers who can validate new and/or improved safety and health techniques.

Defense Variance

The defense variance permits OSHA to grant "reasonable variations, tolerances and exemptions to and from" the Act's requirements "to avoid any serious impairment of the national defense." The defense variance may not be in effect for more than six months without notifying employees and holding a public hearing on the issues.

Multi-state Variances

(Multi-state employers). Employers who operate facilities in more than one state can apply for a multi-state variance. The employer may consolidate the variance application if the employer operates in states that each has its own federally approved OSHA plans. In addition, OSHA permits multi-state employers to use OSHA's variance reciprocity procedures when the request for variance involves State Plan State standards.

Applications for multi-state variances must include:

- A side-by-side comparison of the federal standard and the State Plan State, or those portions of the State Plan State standards that are identical in substance and requirements.
- A certification that the employer has not filed for variance on the same material facts for the same employment or place of employments with any state authority having jurisdiction under an approved OSHA plan.
- A statement with appropriate identification and current status of any citations of violations of the state standards that have been issued to the employer by any of the state authorities and forcing that standard under a plan.

Upon receipt and verification of multi-state variance applications that meet all of the above requirements, federal OSHA furnishes copies to the appropriate State Plan States.

Affected states are provided with the opportunity to comment, and are given the opportunity to participate as a party to the proceedings. State Plan States may review the applications and reach judgments on employers in their states in coordination with federal OSHA. Federal OSHA ensures that each State Plan State involved in the variance determination responds in writing. The written response is made a part of the variance record. If the individual state concurs in the variance, the approval will be reflected in the federal decision. If a state objects to the variance request, it can negotiate separately with the employer to resolve any questions. If the state and employer cannot resolve the differences, the state plan state may make the final determination on that aspect of the variance request under its jurisdiction, independent of the multi-state application.

When a federal OSHA variance has been granted which applies to more than one state (including a state operating under a state plan), the variance becomes the authoritative interpretation of the employers' compliance obligations. This holds true for the federal standard and the identical state standard, except where the individual states have interposed objections.

Additional Employer Rights

As always, additional employer rights are woven into the fabric of all organizations by virtue of their status as business entities. The right to set performance standards and the right to impose discipline govern all aspects of organizational life.

Setting safety performance standards for every position requires that each person work toward these safety goals. Establishing safety performance standards as key criteria in the performance evaluation process holds all employees accountable for meeting their assigned safety responsibilities. In doing so, employees will approach safety as deliberately and seriously as they approach other performance criteria.

Imposing discipline upon any employee who violates company policy, procedures or standards is every employer's right. In those cases when an employee's behavior or actions jeopardize either his own safety or the safety of other employees, the employer has the right and responsibility to take appropriate action.

Intentional or unintentional failure to observe any safety standard must be dealt with promptly and fairly. This is not only in the best interest of the negligent employee, but of other employees and of the company itself.

Finally, employers also have additional rights and responsibilities under the *Occupational Safety and Health Act of 1970* (OSH Act, Public Law 91-596, as amended through January 1, 2004). This holds true for example, once an inspection of the workplace has been conducted by an OSHA compliance safety and health officer (CSHO), in accordance with the OSH Act.

After an OSHA Inspection

After the inspection, the CSHO reports the findings to the Area Director who evaluates them. If a violation exists, OSHA will issue a *Citation and Notification of Penalty* detailing the exact nature of the violation(s) and any associated penalties. A citation informs the company of the alleged violation, sets a proposed time period to correct the violation, and proposes the appropriate dollar penalties.

During the closing conference with the OSHA compliance officer, the compliance officer will discuss or has discussed each apparent violation found during the inspection. The following information can be and should be used as a discussion guide during the company's closing conference with the OSHA compliance officer.

- Nature of the violation;
- Possible abatement measures that may be taken to correct the violation;
- Possible abatement dates that may be required to be met; and
- Any penalties that the Area Director may issue.

The CSHO is a highly trained professional who can help the company/facility recognize and evaluate hazards. The CSHO can also suggest appropriate methods of correcting violations to minimize employee exposure to possible hazardous conditions. In all cases, abatement efforts should always begin as soon as possible.

OSHA Enforcement Activity in the Seven Target Industries Remains Robust

In 2003, OSHA directed their resources towards the reduction of occupational injuries, illnesses, and loss of life. To accomplish these goals, OSHA identified seven industries with high injury/illness rates and a high proportion of severe injuries/illnesses. The focus of this targeting is to accomplish outreach, education and enforcement activity. These industries include:

Landscaping and Horticultural Services
Oil and Gas Field Services
Fruit and Vegetable Processing
Blast Furnace and Basic Steel Products
Ship and Boat Building and Repair
Public Warehousing and Storage
Concrete and Concrete Products

During 2005, OSHA conducted 2,924 inspections within these seven industries. Many of these inspections were a result of Local Emphasis Programs (LEPs), which Area and Regional offices develop to address specific hazards of their geographic location. The objective of this effort is to significantly lower the disproportionately high injury and illness rates in these industries.

Posting Requirements

When a Citation and Notification of Penalty is received, the company owner/facility manager must post the citation, or a copy of it, at or near the place where each violation occurred. Employees, therefore, are aware of the hazards to which they may be exposed. The citation must remain posted for three working days or until the violation is corrected, whichever is longer. *Note:* Saturdays, Sundays, and Federal holidays are not counted as working days. *Posting requirements must be complied with even if the citation is contested.*

Employer Options

The following courses of action are available to the employer who has been cited:

- Agreement with the Citation and Notification of Penalty means that the condition must be corrected by the date set in the citation and the penalty paid, if one is proposed;
- If the employer does not agree, then the employer has fifteen working days from the date the citation is received to contest in writing any of all of the following:
  - citation
  - proposed penalty, and/or
  - abatement date.

Before deciding on either of these options, an informal conference with the OSHA Area Director may be requested. This conference is held to discuss any issues related to the citation and notification of penalty. More information is discussed in the *Informal Conference and Settlement* section that follows later.

OSHA will inform the affected employee representatives of the informal conference or contest.

HOW TO COMPLY

For violations that are not contested, the employer must:

- promptly notify the OSHA Area Director by letter, signed by a member of management, that the appropriate corrective action has been taken within the time set forth in the citation; and

- pay any penalties itemized.

The notification sent to the Area Director is called the Abatement Certification. For Other-Than-Serious violations, this may be a signed letter identifying the inspection number and the citation item number and noting that the violation was corrected by the date specified on the citation. For more serious violations, such as Serious, Willful, Repeat, or failure-to-abate, abatement certification requires more detailed proof.

If the employer has abatement questions after the inspection, the Area Director must ensure that additional information, if available, is provided to the employer as soon as possible.

Employers also can find guidance on abatement verification on OSHA's website at www.osha.gov/Publications/Abate/abate.html.

When the citation permits an extended time for abatement, the employer must ensure that employees are adequately protected during this time. For example, the citation may require the immediate use of personal protective equipment by employees while engineering controls are being installed. When such is the case and where indicated on the citation, the employer must also provide OSHA with an abatement plan, i.e., the steps that will be taken to protect employees and correct the hazards. OSHA must also be provided with periodic progress reports on the actions taken to correct the hazards.

The penalties itemized on the citation and notification of penalty are payable within fifteen working days of receipt of the penalty notice. If however, the employer contests the citation or penalty in good faith, OSHA will suspend abatement and payment of penalties for those items contested until the Occupational Safety and Health Review Commission, or a higher court, issues a final order or rule. The Review Commission is an independent agency and is *not* a part of the U.S. Department of Labor. The final order of the Commission will either uphold, modify, or eliminate the citations and/or penalties. Penalties for items not contested, however, are still due within fifteen working days.

TYPES OF VIOLATIONS

Willful

A willful violation occurs when an employer knows that a hazardous conditions exists and made no reasonable effort to eliminate it. A

Total Violations Remain High; Willful Violations Gain Significantly

The nation's workplaces in 2005 saw 85,307 violations of OSHA's standards and regulations, amounting to a 9.5 percent increase since 2001. The number of willful violations increased 62 percent over 2004 demonstrating that OSHA's enforcement continues to be strong. OSHA identifies employers who intentionally disregard the law, thereby ensuring that employees are protected from serious hazards. OSHA is targeting its resources on those employers who have the highest injury and illness rates and on worksites where employees are more likely to be injured or killed on the job.

OSHA Violation Statistics	*FY2001*	*FY2002*	*FY2003*	*FY2004*	*FY2005*	*% Change 2001-2005*
Total Violations	77,893	77,633	83,539	86,708	85,307	+ 9.5%
Total Serious Violations	52,180	53,845	59,861	61,666	61,018	+ 16.9%
Total Willful Violations	537	331	404	462	747	+ 39%
Total Repeat Violations	1,872	1,867	2,147	2,360	2350	+ 26%
Total Other-than-Serious	22,776	21,128	20,552	21,705	20,819	- 8.6%

willful violation also occurs when the hazardous condition violated a standard, regulation, or the OSH Act. Penalties range from $5,000 to $70,000 per willful violation.

Serious

A serious violation exists when the workplace hazard could cause injury or illness that would most likely result in death or serious physical harm, unless the employer did not know or could not have known of the violation. OSHA may propose a penalty of up to $7,000 for each violation.

Other-than-serious

Other-than-serious violations have a direct and immediate relationship to the safety and health of the exposed employees. However, it is a situation where the most serious situation injury or illness that would be likely to result from a hazardous condition cannot be reasonably be predicted to cause death or serious physical harm to exposed employees. OSHA may impose a penalty of up to $7.000 for each violation.

De Minimis

De minimis violations are violations that have no direct or immediate relationship to safety or health and do not result in citations or penalties.

Failure to Abate

A failure-to-abate violation exists when the employer has not corrected a violation for which OSHA has issued a citation and the abatement date has passed or is covered under a settlement agreement. A failure-to-abate also exists when the employer has not complied with interim measures involved in a long-term abatement within the time given. OSHA may impose a penalty of up to $7,000 per day for each violation.

Repeated

An employer may be cited for a repeated violation if that employer has been cited previously for a substantially similar condition and the citation has become a final order of the Occupational Safety and Health Review commission. A citation is currently viewed as a repeated violation if it occurs within 3 years either from the date that the earlier citation becomes a final order or from the final abatement date, whichever is later. Repeated violations can bring a civil penalty of up to $70,000 for each violation.

For purposes of determining whether a violation is repeated, the following criteria generally apply:

Fixed Establishments

Citations issued to employers having fixed establishments such as factories, terminals and stores that are not normally limited to the cited establishment. A multi-facility employer, for example, can be cited for a repeated violation if the violation recurred at any plant in the nation, and if a citation is obtained and reveals a repeated violation.

Non-fixed Establishments

When an employer is engaged in businesses having no fixed establishments such as construction sites and oil and gas drilling sites, repeated violations are based on employer history and on prior isolations occurring anywhere, and at any of the employer's identified establishments nationwide.

Long-shoring Establishments

A long-shoring establishment covers all long-shoring activities of a single stevedore within any single port area. Long-shoring employers are subject to repeated violation citations based on prior violations occurring anywhere in the nation.

Other Maritime Establishments

Other maritime establishment covered by OSHA standards such as shipbuilding and ship repairing are generally defined as fixed establishments as defined above.

> A violation can be cited as repeated if the employer has been cited for the same or a substantially similar violation anywhere in the nation within the past three years.

Informal Conference and Settlement

Before deciding whether to file a *Notice of Intent to Contest*, the employer may request an informal conference with the OSHA Area Director to discuss the citation and notification of penalty.

The employer may use this opportunity to do any of the following:

- Obtain a better explanation of the violations cited;
- Obtain a more complete understanding of the specific standards that apply;
- Negotiate and enter into an informal settlement agreement;
- Discuss ways to correct violations;
- Discuss problems concerning the abatement dates;

- Discuss problems concerning employee safety practices;
- Resolve dispute citations and penalties, thereby eliminating the need for the more formal procedures associated with litigation before the Review Commission; and
- Obtain answers to any other questions.

OSHA encourages employers to take advantage of the opportunity to have an informal conference if the employer foresees any difficulties in complying with any part of the citation. *Note: An informal conference must be held within the 15-working-day Notice of Intent to Contest period and will neither extend the 15 -working-day contest period nor take the place of the filing of a written notice if the employer desires to contest.* Employee representative(s) have the right to participate in any informal conference or negotiations between the Regional Administrator or the Area Director and the employer.

If the employer agrees that the cited violations exist, but has a valid reason for wishing to extend the abatement date(s), the employer may discuss this with the Area Director in an informal conference. The Area Director then may issue an amended citation that changes the abatement date prior to the expiration of the 15-working-day period without filing a *Notice of Intent to Contest.*

If the citation is not contested within fifteen working days, the citation will become a final order not subject to review by any court or agency. After this occurs, the OSHA Area Director may continue to provide the employer with information and assistance on how to abate the hazards cited, but may not amend or change any citation or penalty which has become a final order. The Area Director may only advise the employer on abatement methods or extend the time needed to abate the violation.

Whenever the employer, an affected employee, or employee representative requests an informal conference, the parties shall be afforded the opportunity to participate fully. If either party chooses not to participate in the informal conference, that party forfeits the right to be consulted before decisions are made that affect the citations. If the requesting party objects to the attendance of the other party, OSHA may hold separate informal conferences. During a joint informal conference, separate or private discussions will be permitted if either party requests them. Informal conferences may be held using any means practical.

HOW TO CONTEST CITATIONS

If an employer wishes to contest any portion of the citation, a *Notice of Intent to Contest* must be submitted in writing within fifteen working days after the receipt of the citation and notification of penalty. This applies even if the employer has stated the disagreement with a citation, penalty or abatement date during a telephone conversation or an informal conference.

The *Notice of Intent to Contest* must clearly state what is being contested—the citation, the penalty, the abatement date, or any combination of these factors. In addition the notice must state whether all the violations on the citation, or just specific violations, are being contested. Example: "Company ABC wishes to contest the citation and penalty proposed for items 3 and 4 of the citation issued December 14, 2006."

The contest, or dispute, must be made in good faith. OSHA will not consider a contest filed solely to avoid a company's responsibilities for abatement or payment of penalties to be a good faith contest.

A proper contest of any item suspends the company's legal obligation to abate and pay until the item contested has been resolved. If the company contests only the penalty, violations indicated on the citation must be corrected by the date indicated. If only some items on the citation are contested, the other items must be corrected by the abatement date and corresponding penalties by paid within fifteen days of notification.

After a *Notice of Intent to Contest* has been filed, the case is then officially in litigation. In order to settle the case, the company may contact the OSHA Area Director who will then direct the company to the attorney handling the case for OSHA. All settlements of contested cases are negotiated between the company and the attorney according to the rules of procedure of the Occupational Safety and Health Review Commission.

The Contest Process

If the company files the written *Notice of Intent to Contest* within the required fifteen working days, the OSHA Area Director forwards the case to the Occupational Safety and Health Review Commission. The Commission assigns the case to an administrative law judge who usually will schedule a hearing in a public place close to the company's workplace. Both employers and employees have the right to participate in this hearing, which contains all the elements of a trial, including examination and cross-examination of witnesses. The company may choose to represent itself or have an attorney represent the company. The administrative law judge may affirm, modify or eliminate any contested items of the citation or penalty.

As with any other legal procedure, there is an appeals process. Once the administrative law judge has ruled, any party to the case may request a further review by the full Review Commission. In addition, any of the three commissioners may bring the case before the entire Commission for review. The commission's ruling, in turn, may be appealed to the U.S. Court of Appeals for the Federal circuit in which the case arose or for the circuit where the employer has its principal office.

Petition for Modification of Abatement

OSHA assigns abatement dates on the basis of the best information available when issuing the citation. If a company is unable to meet an abatement date because of uncontrollable events or other circumstances, and the 15-working-day contest period has expire, a *Petition for Modification of Abatement (PMA)* may be filed with the OSHA Area Director.

The petition must in writing and must be submitted as soon as possible, but no later than one working day after the abatement date. To show clearly that a company has made a good faith effort to comply, the PMA must include all of the following information before OSHA considers it:

- Steps that have been taken to achieve compliance, and the dates they were taken;
- Additional time needed to comply;
- Why additional time is needed;
- Interim steps taken to safeguard the employees against the cited hazard(s) until the abatement; and
- A certification that the petition has been posted, the date of posting and, when appropriate, a statement that the petition has been furnished to an authorized representative of the affected employees. The petition must remain posted for ten working days, during which employees may file an objection.

The OSHA Area Director may grant or oppose a PMA. If it is opposed, it automatically becomes a contested case before the Review Commission. If a PMA is granted, OSHA may conduct a monitoring inspection to ensure that conditions are as they have been described and that adequate progress has been made toward abatement. The OSHA Area Office may provide additional information on PMAs.

What Employees Can Do

Employees or their authorized representatives may contest any or all of the abatement dates set for violations if they believe them to be unreasonable. A written *Notice of Intent to Contest* must be filed with the OSHA Area Director within fifteen working days after the employer receives the citation.

The filing of an employee contest does not suspend the employer's obligation to abate.

Employees also have the right to object to a PMA. Such objections must be in writing and must be sent to the Area Office within ten days of service or posting. OSHA will not make a decision regarding PMA until the Review Commission resolves the issue.

Follow-up Inspections and Failure to Abate

If a citation is received, a follow-up inspection may be conducted to verify the company has done the following:

- Posted the citation as required;
- Corrected the violations as required in the citation; and/or
- Protected employees adequately and made appropriate progress in correcting hazards during multi-step or lengthy abatement periods.

In addition to providing for penalties for failure-to-post citations and failure-to-abate violations, the OSH Act clearly states that employers have a continuing responsibility to comply with the OSH Act and assure the company's employees safe and healthful working conditions. OSHA will cite any new violations discovered during a follow-up inspection.

Employer Discrimination

To achieve abatement by the date set forth in the citation, employers must initiate abatement efforts promptly.

The OSH Act prohibits employers from discharging or otherwise discriminating against an employee who has exercised any right under this law, including the right to make safety and health complaints or to request an OSHA inspection. OSHA will investigate complaints from employees who believe that they have been discriminated against. If the investigation discloses probable violations of employee rights, court action may follow.

Employees who believe that they have been discriminated against must file their complaints within *thirty days* of the alleged act of discrimination. (N.B., additional information about Section 11(c) procedures can be obtained from OSHA).

Providing False Information

All information that employers and employees report to OSHA must be accurate and truthful. Providing false information on efforts to abate cited conditions or in required records is punishable under the OSH Act.

OSHA-Approved State Plans

Currently, 26 states and territories operate OSHA-approved state plans; 23 of these plans cover the private and public sectors and 3 states—Connecticut, New Jersey and New York—cover only public sector employment. For more information, employers and employees in these 26 states and territories should check with their state agencies. State plans may include standards, regulations, and procedures that, while at least as effective as their Federal equivalents, are not always identical to them. For example:

- Some states have different options and procedures for the employer who believes changes, modifications, or deletions of the penalty, citation, or abatement dates are needed.

- Although Federal OSHA *recommends* that employers in general industry, shipbuilding/repair, and marine terminal and long-shoring operations establish comprehensive workplace safety and health programs, some states *require* such programs; and

- In states with OSHA-approved safety and health plans, an employee who believes that he/she has been discriminated against pursuant to Section 11(c) of the OSH Act is entitled to file a complaint alleging discrimination under both state and Federal procedures.

OSHA Assistance

OSHA can provide extensive help through a variety of programs, including technical assistance about effective safety and health programs, state laws, workplace consultations, voluntary protection programs, strategic partnerships, training and education, and more. An overview of the OSHA Assistance programs is found in Appendix I.

WORKER RIGHTS UNDER THE OSH ACT

The law encourages workers to be active players in their workplace's safety and health effort. It gives employees the right to:

- Review copies of appropriate standards, rules, regulations and requirements that the employer is required to have available at the workplace;
- Request information from the employer on safety and health hazards in the workplace, appropriate precautions to take, and procedures to follow if the employee is involved in an accident or is exposed to toxic substances;
- Gain access to relevant employee exposure and medical records;
- Request an OSHA inspection if they believe hazardous conditions or violations of standards exist in the workplace;
- Accompany an OSHA compliance officer during the inspection tour, or have an authorized employee representative do so;
- Respond to questions from the OSHA compliance officer;
- Observe any monitoring or measuring of hazardous materials and see the resulting records, as specified under the OSH Act and required by OSHA standards;
- Review or have an authorized representative review the employer's Log of Work-Related Occupational Injuries and Illnesses (OSHA 300) at a reasonable time and in a reasonable manner;

- Object to the timeframe set by OSHA for the employer to correct a violation by writing to the OSHA Area Director within 15 working days from the date the employer receives the citation;

- Submit a written request to the National Institute for Occupational Safety and Health for information on whether any substance in the workplace has potentially toxic effects in the concentration being used, and, if requested, have their names withheld from the employer;

- Be notified if the employer applies for a variance from an OSHA standard, and have an opportunity to testify at a variance hearing and appeal the final decision;

- Have their names withheld from their employer, by request to OSHA, if they sign and file a written complaint;

- Be advised of OSHA actions regarding a complaint, and request an informal review of any decision not to inspect the site or issue a citation; and

- File a complaint if punished or discriminated against for acting as a "whistleblower" under the OSH Act or 13 other federal statues for which OSHA has jurisdiction; for refusing to work when faced with imminent danger of death or serious injury and there is insufficient time for OSHA to inspect.

Worker Rights to Information

Employer Responsibilities

Employers have a legal obligation to inform employees of OSHA safety and health standards that apply to their workplace. Upon request, the employer must make available copies of those standards and the OSH Act. The employer also must prominently display the official OSHA poster that describes rights and responsibilities under the OSH Act.

Protecting Employees Who Work with Hazardous Materials

Employers must establish a written, comprehensive hazard communication program to ensure that employees who work with or near hazardous materials are informed of the hazards and provided proper protection. A hazard communication program includes provisions for container labeling, material safety data sheets, and an employee training programs. The program must include:

- A list of the hazardous chemicals in each workplace and material safety data sheets for each;
- Methods the employer uses to inform employees of the hazards of non-routine tasks (for example, the cleaning of reactor vessels) and the hazards associated with chemicals in unlabeled pipes in their work areas; and
- A description of methods the employer at a multi-employer work-site will use to inform other employers at the site of the hazards to which their employees may be exposed.

Employee Rights When an Employer Files a Variance

Some employers may not be able to comply fully with a new safety and health standards in the time provided due to shortages of personnel, materials, or equipment. In these situations, employers may apply to OSHA for a temporary variance from the standard. In other cases, employers may prefer to use methods or equipment that differ from those prescribed by OSHA, but which the employer believes are equal to or better than OSHA's requirements. In these cases, the employer may seek a permanent variance for the alternative approach.

The employer's application for a permanent or temporary variance must include certification that:

- The employer has informed workers of the variance application.
- The employee representative receives a copy of the variance application; and
- The employer has posted a summary of the application wherever notices are normally posted in the workplace.

Employers also must inform employees that they have the right to request a hearing on the application. OSHA encourages employees, employers, and other interested groups to participate in the variance process. Notices of variance applications are published in the Federal Register inviting all interested parties to comment on the action.

Worker Rights to Access Records and Test Results

Access to Exposure and Medical Record

Employers must inform employees of the existence, location and availability of their medical and exposure records when they begin employment and then at least annually.

Employers also must provide these records to employees of their designated representatives upon request.

Whenever an employer plans to stop doing business and there is no successor employer to receive and maintain these records, the employer must notify employees of their right of access to these records at least three months before closing the business.

Right to Observe Monitoring Procedures and See Testing Results

OSHA standards require the employer to measure exposure to harmful substances. The employee (or employee representative) has the right to observe the testing and examine the records of the results. If the exposure levels are above the limit set by an OSHA standard, the employer must tell employees what will be done to reduce the exposure.

During an OSHA inspection, an OSHA industrial hygienist may conduct exposure tests if health hazards may be present in the workplace. The inspector may take samples to measure levels of dust, noise, fumes, or other hazardous materials.

OSHA will inform the employee or employee representative who participates in the inspection as to whether the employer is in compliance with OSHA standards. The inspector also will gather detailed information about the employer's efforts to control health hazards, including results of test the employer may have conducted.

Right to Review Injury and Illness Records

An employer with more than 10 employees must maintain records of all work-related injuries and illnesses, and the employees or their representative have the right to review those records. Some industries with very low injury rates are exempt from these recordkeeping requirements.

Worker Rights to Promote Workplace Safety

Working Cooperatively to Reduce Hazards

OSHA encourages employers and employees to work together to reduce hazards. Employees should discuss safety and health problems with the employer, other workers, and, if a labor union exists, union representatives. The OSHA area office can provide information on OSHA requirements. If the worksite is in a state with its own OSHA-approved occupational safety and health program, the state can provide similar information.

Right to Refuse to Perform Unsafe Work

Although nothing in the OSHA law specifically gives an employee the right to refuse to perform an unsafe or unhealthful job assignment, OSHA's regulations, which have been upheld by the U.S. Supreme Court, provide that an employee may refuse to work when faced with an imminent danger of death or serious injury. The conditions necessary to justify a work refusal are very stringent, however, and a work refusal should be taken only as a last resort. If time permits, the employee should report the unhealthful or unsafe condition to OSHA or another appropriate regulatory agency.

Recourse if the Employer Does Not Correct a Hazard

An employee may file a complaint by phone, mail, email, or fax with the nearest OSHA office and request an inspection if there are unsafe or unhealthful working conditions. When doing so, the employee may request that OSHA not reveal his name. If the OSHA area or state office determines that there are reasonable grounds for believing that a violation or danger exists, the office will investigate.

Worker Rights During the Inspection Process

Right to Representation

The OSH Act gives employees or a workers' representative the right to accompany an OSHA compliance officer during an inspection. The OSHA compliance officer is also referred to as a compliance safety and health officer, CSHO, or inspector. The labor union, if one exists, or the employees must choose the representative. The employer, *under no circumstances,* may choose the workers' representative.

If more than one union represents the employees, each union may choose a representative. Normally, union representatives will accompany the inspector in the areas of the facility where their members work. An OSHA inspector may conduct a comprehensive inspection of the entire workplace or a partial inspection limited to certain areas or aspects of the operation.

Right to Help the Compliance Officer

Workers have the right to talk privately to the compliance officer on a confidential basis, whether or not a worker's representative has been chosen. Workers are encouraged to:

- Point out hazards;
- Describe accidents or illnesses that result from those hazards;
- Discuss past worker complaints about hazards; and
- Inform the inspector if working conditions are not normal during the inspection.

Rights to Information Following the Inspection

At the end of the inspection, the OSHA inspector will meet with the employer and the employee representatives in a closing conference to discuss how any hazards that may have been found will be abated. If it is not practical to hold a joint conference, the compliance officer will hold separate conferences. OSHA will provide written summaries, on request.

How to Challenge the Abatement Period

Whether or not the employer accepts OSHA's findings, the employee, or representative, has the right to contest the time OSHA allows for correcting a hazard. This contest, or dispute, must be filed in writing with the OSHA Area Director within 15 working days after the citation is issued. The Occupational Safety and Health Review Commission, an independent agency that is not part of the Department of Labor, will decide whether to change the abatement period.

Right to Information if No Inspection is Conducted or No Citation Issued

The OSHA Area Director evaluates complaints from employees or their representatives and decides whether they are valid. If the area director decides not to inspect the workplace, a certified letter will be sent to the complainant explaining the decision and the reasons for non-inspection.

OSHA will inform complainants that they have the right to request further clarification of the decision from the OSHA Area Director. If still dissatisfied, an appeal to the OSHA regional administrator for an informal review can be made. Similarly, in the event that OSHA decides not to issue a citation after an inspection, employees have a right to further clarification from the Area Director and informal review by the regional administer.

Worker Rights to Protection From Retaliation

Right to Confidentiality

Employees who make a complaint to OSHA about safety and health hazards in their workplaces have a right to confidentiality. If the employee requests that his or her name not be used, OSHA will not tell the employer who filed the complaint or requested an inspection.

Whistleblower Protections

Employees have a right to seek safety and health on the job without fear of punishment. That right is spelled out in Section 11(c) of the OSH Act. The law forbids the employer from punishing or discriminating against employees for exercising such rights as:

- Complaining to the employer, union, OSHA, or any other government agency about job safety and health hazards; and

- Participating in OSHA inspections, conferences, hearings, or other OSHA-related activities.

To protect employees from discharge or discrimination, those states administering their own occupational safety and health programs must have provisions at least as effective as those in the OSH Act. OSHA, however, retains its whistleblower protection authority in all states regardless of the existence of an OSHA-approved state occupational safety and health program.

Workers who believe that have been punished for exercising safety and health rights must contact the nearest OSHA office within 30 days from the time they learn of the alleged discrimination. A representative of the employee's choosing can file the complaint for the worker. Following a complaint, OSHA will contact the complainant and conduct an in-depth interview to determine whether an investigation is necessary.

If the evidence shows that the employee has been punished for exercising safety and health rights, OSHA will ask the employer to restore that worker's job, earnings, and benefits. If the employer refuses, OSHA may take the employer to court. In such cases, a Department of Labor attorney will represent the employee to obtain this relief.

Additional Whistleblower Protection

Since the passage of the OSH Act in 1970, Congress has expanded OSHA's whistleblower protection authority to protect workers from discrimination under 13 additional federal statutes. The agency's investigators receive about 2,000 complaints a year from employees who charge their employer with retaliation. Complaints must be reported to OSHA within set timeframes following the discriminatory action, as prescribed by each law.

Statute Timeframes

These statutes, and the number of days employees have to file a compliant, are:

- ***Occupational Safety and Health Act of 1970 (30 days)***—Provides discrimination protection for employees who exercise a variety of rights guaranteed under the Act, such as filing a safety and health complaint with OSHA and participating in an inspection.

- ***Surface Transportation Assistance Act (180 days)***—Provides protection for truck drivers and other employees relating to the safety of commercial motor vehicles. Coverage includes all buses for hire and freight trucks with a gross vehicle weight greater than 10,001 pounds.

- ***Asbestos Hazard Emergency Response Act (90 days)***—Provides protection for individuals who report violations of environmental laws relating to asbestos in elementary and secondary school systems, public or private.

- ***International Safety Container Act (60 days)***—Protection is provided for employees who report violations of the Act especially those provisions which regulate shipping containers.

- ***Energy Reorganization Act (180 days)***—Provides protection for employees of operators and subcontractors of nuclear power plants licensed by the Nuclear Regulatory Commission. Protection from such discrimination is also provided for employees of contractors working under contract with the Department of Energy.

- ***Clean Air Act (30 days)***—Protection is provided for employees who report violations of the Act, regarding air quality and air pollution.

- ***Safe Drinking Water Act (30 days)***—Provides protection for employees who report violations of the Act, which requires that all drinking water systems in public buildings and new construction of all types be lead free.

- ***Federal Water Pollution Control Act (30 days)***—Provides protection for employees who report violations of regulations involving the manufacture, distribution, and use of certain toxic substance.

- ***Toxic Substances Control Act (30 days)***—Provides protection for employees who report violations of regulations involving the manufacture, distribution, and use of certain toxic substances.

- ***Solid Waste Disposal Act (30 days)***—Also called the *Resource Conservation and Recovery Act*. Protection is given to employees who exercise certain rights under the Act. This Act provides assistance for the development of facilities for the recovery of energy and other resources from discarded materials. This Act also regulates hazardous waste management.

- ***Comprehensive Environmental Response, Compensation, and Liability Act (30 days)***—Provides protection for employees who exercise rights under the Act. This Act provides liability, compensation, cleanup, and emergency response for hazardous substances released into the environment and for the cleanup of inactive hazardous waste disposal sites.

- ***Wendell H. Ford Aviation Investment and Reform Act for the 21st Century (90 days)***—Provides protection for employees of air carriers, contractors, or subcontractors of air carriers who raise safety concerns.

- *Corporate and Criminal Fraud Accountability Act of 2002 (90 days)*—Provides protection for employees of publicly traded companies or brokerage firms or their contractors, subcontractors, or agents who report violations of the Act. This Act covers mail, wire, bank, or securities fraud or violations of laws related to fraud against stockholders.

- *Pipeline Safety Improvement Act of 2002 (180 days)*—Provides protection for employees who report violations of the federal law regarding pipeline safety and security or who refuse to violate such provisions.

A further discussion of the whistleblower act will be found in Chapter 3.

Worker Rights in State-Plan States

States that assume responsibility for their own occupational safety and health programs must have provisions at least as effective as Federal OSHA's, including the protection of employee rights.

Any interested person or group, including employees, with a complaint concerning operation or administration of a state plan may submit a complaint to the appropriate OSHA regional administrator. (See Appendix I for contact information). The OSHA regional administrator will investigate the complaints and inform the state and the complainant of these findings. When appropriate, OSHA will recommend corrective action.

Worker Responsibilities

Although OSHA does not cite employees for violations, the requires that each employee "shall comply with all occupational safety and health standards and all rules, regulations, and orders issued under the Act" that are applicable. Each employee should:

- Read the OSHA poster at the jobsites.

- Comply with all applicable OSHA standards.

- Follow all lawful employer safety and health rules and regulations, and wear or use prescribed protective equipment while working.

- Report hazardous conditions to the supervisor.
- Report any job-related injury or illness to the employer, and seek treatment promptly.
- Cooperate with the OSHA compliance officer conducting an inspection if he or she inquires about safety and health conditions in the workplace; and
- Exercise rights under the OSH Act in a responsible manner.

Sources

Gustin, Joseph F., *Safety Management: A Guide For Facility Managers,* New York: UpWord Publishing, Inc., 1996.

Gustin, Joseph F., *Disaster and Recovery Planning: A Guide For Facility Managers, 4th ed.*, Lilburn, GA: The Fairmont Press, Inc., 2007.

The OSH Act of 1970 (December 29, 1970), PL 91-596, 29 USC 651

United States Department of Labor, Occupational Safety and Health Administration, *Employer Rights and Responsibilities Following an OSHA Inspection,* OSHA 3000-08R, 2005.

United States Department of Labor, Occupational Safety and Health Administration, *Employee Workplace Rights,* OSHA 3021-01R, 2005.

Chapter 3

The Whistleblower Act

Employees may file a complaint with OSHA if they are retaliated against by their employer while involved in protected activity relating to workplace safety and health, commercial motor carrier safety, pipeline safety, air carrier safety, nuclear safety, the environment, asbestos in schools, corporate fraud, or SEC rules or regulations.

WHISTLEBLOWER LAWS ENFORCED BY OSHA

Each law requires complaints to be filed within a specified time frame after the alleged retaliation.

Per OSHA, complaints may be filed by telephone or in writing under any of the following Acts:

- Occupational Safety and Health Act (30 days)
- Surface Transportation Assistance Act (180 days)
- Asbestos Hazard Emergency Response Act (90 days)
- International Safe Container Act (60 days)

Under the following laws, complaints must be filed in writing:

- Clean Air Act (30 days)
- Comprehensive Environmental Response, Compensation and Liability Act (30 days)
- Energy Reorganization Act (180 days)
- Federal Water Pollution Control Act (30 days)
- Pipeline Safety Improvement Act (30 days)
- Safe Drinking Water Act (30 days)
- Sarbanes-Oxley Act (90 days)
- Solid Waste Disposal Act (30 days)
- Toxic Substances Control Act (30 days)

- Wendell H. Ford Aviation Investment and Reform Act for the 21st Century (90 days)

Unfavorable Employment Actions

Retaliation by an employer may be found if the protected activity of an employee was a contributing or motivating factor in its decision to take unfavorable employment action against the employee. Such actions may include:

- Discharge or layoff
- Blacklisting
- Demotion
- Denial of overtime or promotion
- Disciplining
- Denial of benefits
- Failure to hire or rehire
- Intimidation
- Reassignment affecting prospects for promotion
- Reduction in pay or hours

Filing a Complaint

If an employee believes that retaliation has taken place because the employee exercised the legal right as an employee, then that employee must contact the local OSHA office as soon as possible because of the legal time limits. Complaints may be telephoned, faxed or mailed to the nearest OSHA office (see Appendix I for a listing of the OSHA offices). OSHA will then conduct an in-depth interview with each complainant to determine whether to conduct an investigation.

If retaliation for protected activity relating to occupational safety and health issues takes place in a state that operates an OSHA-approved state plan, the complaint should be filed with the state agency, although persons in those states may file with Federal OSHA at the same time. Although the Occupational Safety and Health Act covers only private sector employees, state plans also cover state and local government employees.

How OSHA Determines Whether Retaliation Took Place

There are a number of factors that must be substantiated in the investigation to determine if retaliation did, in fact, occur.

The investigation must reveal that:

- The employee engaged in protected activity;
- The employer knew about the protected activity; and
- The protected activity was the motivating factor, or under some laws, a contributing factor in the decision to take adverse action against the employee.

If the evidence supports the employee's allegation and a settlement cannot be reached, OSHA will issue an order requiring the employer to reinstate the employee, pay back wages, restore benefits, and other possible remedies to make the employee whole.

Limited Protections for Employees Who Refuse to Work

An employee has a *limited* right under the OSH Act to refuse to do a job because conditions are hazardous. The employee may do under the OSH Act only when:

- The employee believes that death or serious injury is faced and the situation is so clearly hazardous that any responsible person would believe the same thing;
- The employee has tried to get the employer to correct the condition, and there is no other way to do the job safely; and
- The situation is so urgent that employee does not have time to eliminate the hazard through regulatory channels such as calling OSHA.

Regardless of the unsafe condition, employees are not protected if they simply walk off the job. OSHA cannot enforce union contracts or state laws that give workers the right to refuse to work.

WHISTLEBLOWER PROTECTIONS IN VARIOUS INDUSTRIES

Whistleblower Protections in the Transportation Industry.

Employees whose jobs directly affect commercial motor vehicle safety are protected from retaliation by their employers for the following:

- refusing to violate or for reporting violations of Department of Transportation, (DOT), motor carrier safety standards or regulations;
- refusing to operate a vehicle because of such violations; or
- because they have a reasonable apprehension of death or serious injury.

Similarly, employees of air carriers, their contractors or subcontractors who raise safety concerns or report violations of FAA rules and regulations are protected from retaliation. Additionally, employees of owners and operators of pipelines, their contractors and subcontractors, who report violations of pipeline safety rules and regulations are also protected from retaliation. Employees involved in international shipping who report unsafe shipping containers are also protected.

Whistleblower Protections for Voicing Environmental Concerns

A number of laws protect employees who report violations of environmental laws related to drinking water and water pollution, toxic substances, solid waste disposal, air quality and air pollution, asbestos in schools, and hazardous waste disposal sites. The Energy Reorganization Act protects employees who raise safety concerns in the nuclear power industry and in nuclear medicine.

Whistleblower Protections When Reporting Corporate Fraud

Employees who work for publicly traded companies or companies required to file certain reports to the Securities and Exchange Commission are protected from retaliation for reporting alleged violations of mail, wire, or bank fraud; violations of rules or regulations of the SEC, or federal laws relating to fraud against shareholders.

Since January, 2002, OSHA has been delegated responsibility for two additional whistleblower statutes, the Corporate and Criminal Fraud Accountability Act of 2002 and the Pipeline Safety Improvement Act of 2002.

THE SARBANES-OXLEY ACT

Filing Whistleblower Complaints under the Sarbanes-Oxley Act

A company is covered by section 806 of the Sarbanes-Oxley Act of 2002 (the Act) if it has a class of securities registered under Section 12 of the Securities Exchange Act, or is required to file reports under Section 15(d) of that Act. Its contractors, subcontracts, or agents may also be covered.

Protected Activity

If an employer is covered under the Sarbanes-Oxley Act, it may not discharge or in any manner retaliate against an employee because that employee may have:

- provided information;
- caused information to be provided, or
- assisted in an investigation by
  - a federal regulatory or law enforcement agency;
  - a member or committee of Congress, or
  - an internal investigation by the company relating to an alleged violation of mail fraud, wire fraud, bank fraud, securities fraud, or violating SEC rules or regulations or federal laws relating to fraud against shareholders.

In addition, an employer may not discharge or in any manner retaliate against employees because they may have filed, caused to be filed, or participated in or assisted in a proceeding under on of these laws or regulations.

If an employer takes retaliatory action against an employee because he or she engaged in any of these protected activities, the employee can file a complaint with OSHA.

Unfavorable Employment Actions

An employer may be found to have violated the Sarbanes-Oxley Act if an employee's protected activity was a contributing factor in the employer's decision to take unfavorable employment action against the employee. Such actions may include:

- Discharge or layoff
- Blacklisting
- Demoting
- Denial of overtime or promotion

- Disciplining
- Denial of benefits
- Failure to hire or rehire
- Intimidation
- Reassignment affecting prospects for promotion
- Reduction in pay or hours

Deadline for Filing Complaints

Complaints must be filed within 90 days after an alleged violation of the Act (Sarbanes-Oxley) occurs; i.e., when the complainant becomes aware of the retaliatory action. An employee, or representative of an employee, who believes that they have been retaliated against in violation of the Act may file a complaint with OSHA.

How to File a Complaint

Complaints must be filed in writing. The complaint should be filed with the OSHA office responsible for enforcement activities in the geographic area where the employee resides or was employed, but may be filed with any OSHA officer or employee.

Complaints may be filed by mail; certified mail is recommended. Complaints may also be faxed or hand delivered during business hours. The date postmarked, faxed or hand delivered is considered the date filed.

Contents of a Complaint

There is no set form for the complaint, but it must be filed in writing and include the following:

- The name, address and phone number(s) of the person filing the complaint, or on whose behalf the complaint is being filed, must be included.

- The names and addresses of the company(s) and person(s) who are alleged to have violated the Act—who the complaint is being filed against.

- Sufficient detail to allege the four elements of a *prima facie* violation:

— The employee engaged in a protected activity or conduct;

— The employer or named person knew or suspected, actually or constructively, that the employee engaged in the protected activity;

— The employee suffered an unfavorable personnel action; and

— The circumstances were sufficient to raise the inference that the protected activity was a contributing factor in the unfavorable action.

When it receives the complaint, OSHA will review it to determine whether to conduct an investigation, and will conduct its investigation in accordance with the Act's requirements.

Results of the Investigation

If the evidence supports an employee's claim of retaliation and a settlement cannot be reached, OSHA will issue an order requiring the employer to reinstate the employee, pay back wages, restore benefits, and other possible relief to make the employee whole, including:

- Reinstatement with the same seniority status.
- Payment of back pay with interest.
- Compensation for special damages, attorney's fees, expert witness fees, and litigation costs.

OSHA's findings and order become the final order of the Secretary of Labor, unless they are appealed within 30 days.

After OSHA issues its findings and order, either party may request a full hearing before an administrative law judge of the Department of Labor. The administrative law judge's decision and order may be appealed to the Department's Administrative Review Board for review.

If a final agency order is not issued within 180 days from the date the employee's complaint is filed, then the employee may file it in the appropriate United States district court.

Sources

U.S. Department of Labor, Occupational Safety and Health Administration, Directive Number: CPL 02-03-002—*Whistleblower Investigations Manual,* August, 2003.

U.S. Department of Labor, Occupational Safety and Health Administration, OSHA Fact Sheet, "Filing Whistleblower Complaints under the Sarbanes-Oxley Act," December, 2006.

U.S. Department of Labor, Occupational Safety and Health Administration, OSHA Fact Sheet, "Your Rights as a Whistleblower," December, 2006.

Chapter 4

Safety Regulations and Facility Liability

Poor safety practices can result in many direct and hidden costs to the employer's bottom line—including penalties for non-compliance with OSHA standards and the higher cost of insurance premiums. For these reasons, it pays for facility managers to know how OSHA makes its standards, what safety and health regulations are on the books and are in process of change, and key liability issues.

OSHA STANDARDS

While the process for establishing safety standards follows a basic pattern, there are, obviously, safety issues and hazards that are unique to a particular industry.

OSHA addresses these differences by issuing two types of standards: Horizontal and Vertical.

Most are Horizontal Standards, also known as General Industry Standards. Horizontal Standards cover hazards that are common to a wide variety of industries.

Vertical standards apply solely to one industry. OSHA has promulgated Vertical Standards for the construction, agriculture and maritime industries. Note that some Horizontal Standards also apply to these industries as well.

Regardless of their classification as Horizontal or Vertical, OSHA standards are designed to reduce on-the-job injuries and to limit or minimize workers' risk of developing occupational disease.

THE STANDARD-SETTING PROCESS

Standards-setting is a thorough, detailed and time-consuming process that involves data collection/analysis, negotiation, public comment and, as is often the case, public hearings.

As needs arise and additional information—such as scientific, medical and injury/illness statistical data—becomes available, the standard-setting process is set into motion. The process may be initiated by OSHA, Congress, NIOSH, the U.S. Environmental Protection Agency's (EPA) Toxic Substance Control Act (TOSCA) referral, any OSHA Advisory Committee or through public petition.

Standard-setting begins with publication in the *Federal Register* of a Request for Information (RFI), an Advance Notice of Proposed Rulemaking (ANPR), or a Notice of Proposed Rulemaking (NPRM). Through the RFI or the ANPR, OSHA actively seeks information to determine the extent of a particular hazard, protective measures that are currently being used (as well as potential measures) and cost and befits of various protective strategies.

Negotiated Rulemaking

OSHA begins the standard-setting process by developing consensus through negotiated rulemaking.

The agency forms an advisory committee comprised of industry and labor representatives whose constituents will be affected by the standard. These representatives work out an agreement that serves as the basis for the proposed rule.

Negotiated rulemaking shortens the time frames for rulemaking, discourages legal challenges to the final rule, and provides opportunity for full public comment on the issue.

The information gathered serves as the basis for developing a proposal. At times, OSHA circulates preliminary drafts of proposals for informal comment from any of the affected industry and labor groups. A public comment period of 60-90 days usually follows a formal proposal's publication in the *Federal Register*, and may be extended at the request of any interested party.

In addition, any interested party or persons may request public hearings on a proposal. While Department of Law administrative judges preside over public hearings, all decisions regarding the final standard are made by OSHA. Public hearings are followed by post-hearing comment periods, usually 30 or more days.

The final step in the rulemaking process occurs when OSHA publishes either the final standard or the determination that the standard is not needed. Publication in the *Federal Register* also sets the effective date for the implementation of the standard, which normally occurs 90 or fewer days from the standard's issuance date. However, depending upon the detailed provisions of either the standard or the engineering controls required to implement the standard, the standard may be phased in over a longer period.

Any final OSHA standard may be challenged in the appropriate U.S. Circuit Court of Appeals by any party that is adversely affected by the final standard.

Health Standards: Special Requirements

In establishing health standards, OSHA follows a four-step process. First, OSHA must demonstrate that a hazard, or potential hazard, poses a significant risk to workers' health.

Next, the agency must show that a standard would either eliminate the risk, or substantially reduce the risk.

OSHA must then select the most protective exposure limit(s) that is both technologically and economically feasible.

Finally, OSHA must look for the cost-effective method(s) for employers to meet the exposure limit.

Standard Setting: Determining Priorities

The U.S. Department of Labor publishes a semiannual agenda of regulatory activity in the *Federal Register.*

The agenda covers anticipated activity for the year following publication, including the standards that are currently being developed and target dates for compliance.

Emergency Temporary Standards

Emergency temporary standards are established by OSHA when the agency determines that a "grave danger" exists for workers who are exposed to toxic substances or to any other physical conditions that might by similarly hazardous. They are issued for a six-month period. However, they remain in effect until permanent standards are developed by OSHA.

State Standards

State Plan States' standards must be "at least as effective" as federal standards. These equally effective state standards must be issued within six months of publication of the federal OSHA standards in the

Federal Register

State Plan States can develop and issue standards that are not regulated by federal OSHA. When applicable to products that are distributed or used in interstate commerce, these state standards must be "required by compelling local conditions: and must not "unduly burden interstate commerce."

Variances

As previously noted, variances provide employers with alternatives to compliance standards. A ***temporary variance*** allows employers a short-term exemption from a standard when they cannot comply with OSHA requirements by the prescribed date. This exception is given when the necessary construction or alteration of the facility cannot be completed in time or when technical personnel, materials, or equipment are temporarily unavailable.

To be eligible for a temporary variance, an employer must implement an effective compliance program as quickly as possible. In the meantime, the employer must demonstrate to OSHA that all available steps are being taken to safeguard employees. Inability to afford compliance costs is not a valid reason for requesting a temporary variance. Details on the information required in the application for a temporary variance are found in 29 CFR Part 1905.10 Subpart B.

Temporary variances are granted by OSHA when an employer successfully demonstrates that additional time beyond the effective date is needed to company with a newly issued standard.

An ***interim order*** permits an employer to continue operating under existing conditions while OSHA considers the employer's request for a variance—whether a temporary variance, a permanent variance, or both. If the interim order is granted, employers must inform workers by giving a copy to authorized employee representatives and posting a copy prominently in the workplace.

A ***permanent variance*** grants an exemption from a standard to an employer who can prove that their methods, conditions, practices, operations, or processes provide workplaces that are as safe and healthful as those that follow the OSHA standards. When applying for a permanent variance, employers must let employees know they have filed the application and that the employees have the right to request a hearing.

To decide whether to grant a permanent variance, OSHA reviews the employer's evidence and, if appropriate, visits the workplace to confirm the facts provided in the application. If the request has merit, OSHA may grant a permanent variance. Final variance orders detail the employer's specific responsibilities and requirements. Final variance also explains exactly how the employer's method varied from the OSHA requirement. Details on the information required in the application for a permanent variance are found in 29 CFR Part 1905.11 Subpart B.

When an employer can satisfactorily demonstrate to OSHA that its proposed alternative will provide a safe and healthful working environment equal to or exceeding the OSHA standard, the agency will grant a permanent variance.

Employers use ***experimental variances*** to demonstrate or validate new or improved safety and health techniques. They are developed in close cooperation with OSHA. Details on the information required in the application for an experimental variance are found in section 6(b)(6)(C) of the Occupational Safety and Health Act (OSH ACT).

OSHA uses ***national defense variances*** to grant "reasonable variations, tolerances and exemptions to and from" the requirements of the OSH Act "to avoid serious impairment of the national defense." If a defense variance is in effect for more than six months, employer must notify employees and offer a public hearing on the issues involved. Details on the information required in the application for a national defense variance are located in 29 CFR Part 1905.12 Subpart B.

Application for a Variance

Application for a variance is dependent upon the jurisdiction of the worksite. If the worksite involved is under federal jurisdiction, then variance applications are forwarded to the U.S. Department of Labor/OSHA Office of Technical Programs and Coordination Activities, in Washington, DC. States and territories included in federal jurisdiction are:

- Alabama
- Arkansas
- Colorado
- DC
- Delaware
- Florida
- Georgia
- Idaho
- Illinois
- Louisiana
- Massachusetts
- Maine
- Missouri
- Mississippi
- Montana
- North Dakota
- Nebraska
- New Hampshire
- Oklahoma
- Ohio
- Pennsylvania
- Rhode Island
- South Dakota
- Texas
- Wisconsin
- West Virginia
- American Samoa

- Guam
- Trust Territory of the Pacific Islands
- Wake Island

Employers whose worksites are located in state or territorial jurisdiction should address their variance applications to the appropriate state or territorial OSHA office, located in the State/Territory's Department Of Labor of a state or territory. The following states and territories operate under their own OSHA-approved programs:

- Alaska
- Arizona
- California
- Connecticut
- Hawaii
- Iowa
- Indiana
- Kentucky
- Maryland
- Michigan
- North Carolina
- New Jersey
- New Mexico
- Nevada
- New York
- Oregon
- South Carolina
- Tennessee
- Utah
- Virginia
- Vermont
- Washington
- West Virginia
- Puerto Rico
- Virgin Islands

Note that Connecticut, New Jersey and New York cover only public sector employees.

Multi-State Worksites

If an employer has operations in more than one state, that employer may apply for a multi-state variance. Even if one or more of the states involved has its own OSHA-approved state plan, the employer may consolidate the variance applications. If a state standard is identical to the federal standard, the state will have the opportunity to join with federal OSHA in granting a variance.

SAFETY AND HEALTH REGULATIONS

As new technology is introduced into the mainstream of the U.S. workforce and new information including scientific and health data is learned, safety and health standards are subject to revision and/or amendment. Below are several examples.

Asbestos

In the revised Asbestos Standard, employers must ensure that no employee is exposed to an airborne concentration of asbestos in excess of 0.1 f/cc as an 8-hour time-weighted average (TWA). In addition, employees must not be exposed to an airborne concentration of asbestos in excess of 1 f/cc as averaged over a sampling period of 30 minutes.

If this level is exceeded, employers are required to begin compliance activities, including air monitoring, employee training and medical surveillance.

This standard also requires comprehensive notification regarding the presence, location and quantity of asbestos-containing material. Even though employees may not be at risk of exposure, building and facility owners are required to provide notification. The requirement is based on the presumption that buildings and facilities constructed before 1980 contain asbestos. Building and facility owners must comply with notification procedures, unless they can prove that the pre-1980 construction does not contain asbestos.

Hazard Communication

In 1994, the Hazard Communication Standard was amended to include warnings on shipping container labels. Employers who receive bulk shipping containers, rail cars, motor and/or transport vehicles at the worksite are required to retain the markings required by the U.S. Department of Transportation's (DOT) Hazardous Materials Regulations. These markings must be retained until the bulk shipping containers are sufficiently cleaned of any and/or all residue and purged of vapors to remove any potential hazards. Bulk containers, and those that will be reshipped, must retain all DOT labels, as well as labels that employers are required to affix. Non-bulk packages that are not intended for reshipment do not need to retain the DOT labeling.

With an effective compliance date of March 11, 1993, the standard also requires that distributors of chemicals provide material safety data sheets (MSDS) to "downstream users." The standard also clarifies the definition of "article" to specify that any release of trace amounts of a hazardous chemical will not affect the article exemption, as long as the release does not pose a physical or health hazard to employees.

Fatalities and Multiple Hospitalizations

The reporting requirements for fatalities and multiple hospitalizations were revised in April, 1994. This requirement reduces the time to report any fatalities and multiple hospitalizations from 48 hours to eight hours. The regulation applies to all deaths from work-related incidents and to all hospitalization of three or more workers that result from work-related incidents. Employers are required to report these incidents by telephone, or in person, to the nearest OSHA area office or to the OSHA central number at those times when OSHA area offices are closed.

Information that must be reported includes:

- The employer's name
- The location of the incident.
- A description of the incident, in addition to the actual number of fatalities and/or multiple hospitalizations.
- A contact person and telephone number.

If an employer does not learn of the fatality or multiple hospitalizations within eight hours of occurrence, it must report the incidents within eight hours of learning of the incident(s).

Additionally, the reporting requirements mandate that any fatality or multiple hospitalizations of three or more employees that occurs within 30 days of the work-related incident must be reported to OSHA.

Personal Protective Equipment

The Personal Protective Equipment Standard (PPE) was revised in April, 1994. The purpose for the revision was to update the standard to reflect current technological improvements to personal protective equipment; conduct assessments to determine the presence of actual and potential hazards (including those created by impact, compression, penetration, heat, chemicals, harmful dust and light radiation) which necessitate or would necessitate the use of personal protective equipment; and provide employee training in the use of personal protective equipment.

The PPE now requires that employers certify in writing that they have performed the assessment, and verify in writing that employees have been trained on and understand the necessity, use, limitations and proper care and maintenance of personal protective equipment. The written certification must:

- Identify the worksite evaluated.
- List the dates the assessment was performed.
- Specify that the written document is a certification of the assessment.

The effective compliance dates for the revised PPE were July 5, 1994 for the PPE upgrades, and October 5, 1994 for the hazard assessment and training requirements.

GENERAL INDUSTRY REQUIREMENTS

Literally hundreds of industry-specific and general industry safety and health standards exist. The following are selected OSHA requirements that apply to many general industry employers.

Hazard Communication Standard

This standard is designed to ensure that employers and employees know about hazardous chemicals in the workplace and how to protect themselves. Employers with employees who may be exposed to hazardous chemicals in the workplace must prepare and implement a written Hazard Communication Program and comply with other requirements of the standard.

Emergency Action Plan Standard

OSHA recommends that all employers have an Emergency Action Plan. A plan is mandatory when required by an OSHA standard. An Emergency Action Plan describes the actions employees should take to ensure their safety in a fire or other emergency situation.

Fire Safety

OSHA recommends that all employers have a Fire Prevention Plan. A plan is mandatory when required by an OSHA standard.

Exit Routes

All employers must comply with OSHA's requirements for exit routes in the workplace.

Walking/Working Surfaces

Floors, aisles, platforms, ladders, stairways, and other walking/working surfaces are present, to some extent, in all general industry workplaces. Slips, trips, and falls from these surfaces constitute the majority of general industry accidents. The OSHA standards for walking and working surfaces apply to all permanent places of employment, except where only domestic, mining, or agricultural work is performed.

Medical and First Aid

OSHA requires employers to provide medical and first aid personnel and supplies commensurate with the hazards of the workplace. The details of a workplace medical and first aid program are dependent on the circumstances of each workplace and employer.

These standards are generally required of all workplaces within the industry. Every employer must comply with these requirements or the parallel state plan requirements, except where specifically exempted.

OTHER OSHA REQUIREMENTS

Additionally, a number of other OSHA standards may apply to the workplace. The following checklist can help identify other key OSHA standards that may apply and point to information to help ensure compliance with those standards.

1. Companies that employ machine operators such as those who operate saws, slicers, shears, slitters, ,power presses, etc., may be subject to OSHA's Machine Guarding requirements.

2. If there are employees who service or maintain machines or equipment that could start up unexpectedly or release hazardous energy, the employer may be subject to OSHA's Lockout/Tagout requirements.

3. Electrical hazards, such as wiring deficiencies, are one of the hazards most frequently cited by OSHA. OSHA's electrical standards include design requirements for electrical systems and safety-related work practices.

4. Employers must perform an assessment of each operation in their workplace to determine if their employees are required to wear personal protective equipment (PPE). Note that engineering controls and work practices are the preferred methods for protecting employees—OSHA generally considers PPE to be the least desirable means of controlling employee exposure.

5. If necessary to protect the health of employees, appropriate respirators must be provided. A Respiratory Protection program that meets the requirements of OSHA's Respiratory Protection standard must also be established.

6. Employers whose employees are exposed to excessive noise (e.g., conditions that make normal conversation difficult) may be required to implement a Hearing Conservation program.

7. Employers should evaluate their workplaces for the presence of confined spaces.

8. If employees may be exposed to blood or bodily fluids as part of their assigned duties, employers may be subject to OSHA's Bloodborne Pathogens standard.

9. If employees operate Powered Industrial Trucks (i.e., forklifts), the company may be subject to OSHA's Powered Industrial Trucks standard.

This list is not comprehensive. Additional OSHA standards may apply to a particular workplace. OSHA's general industry standards (29 CFR 1910) should be reviewed for other requirements. In addition, section 5(a)(1) of the Occupational Safety and Health Act, known as the General Duty Clause, requires employers to provide their employees with a workplace that is free of recognized hazards likely to cause death or serious physical harm.

Survey the Workplace for Additional Hazards

Workplaces should be surveyed for additional hazards and OSHA requirements. OSHA's Self-Inspection Checklists (see Appendix III) should be used to conduct a survey.

Develop a Comprehensive Jobsite Safety and Health Program

While OSHA does not require employers to develop comprehensive safety and health programs, development and implementation of these programs is an effective way to comply with OSHA standards and prevent workplace injuries and illnesses.

Recordkeeping, Reporting and Posting

Recordkeeping

OSHA requires certain employers to keep records of workplace injuries and illnesses (29 CFR 1904).

- First employers must determine if they are exempt from the routine recordkeeping requirements. Certain employers not required to keep OSHA injury and illness records unless asked to do so in writing by either OSHA or the Bureau of Labor Statistics. This occurs if the employer had 10 or fewer employees during all of the last calendar year (29 CFR 1904.1); or

- If the employer is in a certain low-hazard retail, service, finance, insurance, or real estate industries (29 CFR Part 1904, Subpart B, Appendix A).

- If an employer does not qualify for these exemptions, compliance with OSHA's recordkeeping requirements is mandatory.

Reporting

OSHA requires all employers, regardless of size or industry, to report the work-related death of any employee or hospitalizations of three or more employees. (29 CFR 1904.39).

OSHA Poster

All employers must post the OSHA Poster (or state plan equivalent) in a prominent location in the workplace.

Access to Employee Exposure and Medical Records

An OSHA standard (29 CFR 1910.1020) requires employers to provide employees, their designated representatives, as well as OSHA, with access to employee exposure and medical records. Employers generally must maintain employee exposure records for 30 years and medical records for the duration of the employee's employment plus 30 years.

NOTE: If the workplace is in a state operating an OSHA-approved state program, state plan recordkeeping regulations, although substantially identical to federal ones, may have some more stringent or supplemental requirements. In such cases, the state program agency should be contacted directly for additional information.

Sources

Gustin, Joseph F., *Safety Management: A Guide For Facility Managers*, New York: UpWord Publishing, Inc., 1996.

Gustin, Joseph F., *Disaster and Recovery Planning: A Guide For Facility Managers, 4th ed.*, Lilburn, GA: The Fairmont Press, Inc., 2007.

The OSH Act of 1970 (December 29, 1970), PL 91-596, 29 USC 651

United States Department of Labor, Occupational Safety and Health Administration, *Asbestos.* OSHA Fact Sheet.

United States Department of Labor, Occupational Safety and Health Administration, *Hazard Communication Standard.* Fact Sheet No. OSHA 93-26.

United States Department of Labor, Occupational Safety and Health Administration, *Model Plans and Programs for the OSHA Bloodborne Pathogens and Hazard Communications Standards.* OSHA 3186-06N, 2003.

United States Department of Labor, Occupational Safety and Health Administration, *OSHA Compliance Assistance, Compliance Assistance Quick Start. www.osha.gov.*

United States Department of Labor, Occupational Safety and Health Administration, *Personal Protective Equipment.* OSHA Fact Sheet.

United States Department of Labor, Occupational Safety and Health Administration, *Variances from OSHA Standards.* OSHA Fact Sheet.

Chapter 5

The Americans with Disabilities Act

Perhaps no other legislative act in recent history has impacted facility management as much as the Americans with Disabilities Act (ADA). Its scope is far reaching and pervades every aspect of facility management. From the employment provisions (Title I) of the ADA, to the Public Accommodations and Commercial Facilities provisions (Title III), the ADA defines, in large part, the role of facility manager as compliance officer.

The areas presented in this discussion of the ADA are those that have particular relevance for facility managers. The significance of the Title I employment provisions cannot be underestimated. Title I determines, in large part, the approach facility managers must take to meet their responsibilities to ensuring worker safety and in managing the safety function.

Title II, which governs state and local government operations, as well as Title III, which covers private entities that have been determined to be places of public accommodation or commercial facilities, carry additional responsibilities for facility managers in both the public and private sectors.

These responsibilities include the facility managers' obligations to maintain environments that ensure the safety of people with disabilities, but also minimize risk of injury to those employees who provide services to people with disabilities.

BACKGROUND

The ADA is a comprehensive civil rights law for people with disabilities. The Equal Employment Opportunity Commission (EEOC) enforces the ADA's Title I employment practices by private entities. The Department of Justice enforces the ADA's requirements in three areas:

Title I: Employment practices by units of state and local governments

Title II: Programs, services and activities of state and local governments

Title III: Public accommodations and commercial facilities of private entities.

Title I: Employment

The ADA was signed into law on July 26, 1990. The employment provisions of the Act, referred to as Title I, became effective on July 26, 1992 for employers with 25 or more employees. July 26, 1994 marked the effective compliance date for employers with 15 to 24 employees.

The ADA prohibits employers from discriminating against qualified individuals with a disability who, with or without reasonable accommodations, can perform the essential functions of a job. This prohibition extends to:

- The job application process.
- Any term of employment, including hiring, advancement, discharge, compensation, job training, and any and all other terms, conditions and privileges of employment.

The intent of the ADA is to ensure access to employment based on merit. The ADA does not create preferences in favor of individuals with disabilities, nor does it establish quotas or affirmative action requirements.

Title II: Programs, Services and Activities of State and Local Government

Title II provisions of the ADA cover "public entities" which include the "state or local government and any of its departments, agencies, or other instrumentalities."

The specific focus of Title II is on any and/or all activities, services and programs of public entities that include the activities of:

- State legislatures and courts.
- Town meetings.

- Police and fire departments.
- Motor vehicle licensing.
- Employment.

Municipally operated public transportation systems, as well as other state and local government-operated transportation systems are covered by Department of Transportation regulations.

These regulations establish specific requirements for transportation vehicles and facilities, including a requirement that all new buses must be equipped to provide services to people who use wheelchairs.

Title III: Public Accommodations and Commercial Facilities

Referred to as Title III of the ADA, the provisions concerning public accommodations and commercial facilities became effective on January 26, 1992. The significance of Title III is twofold:

1. Title III mandates that public accommodations and commercial facilities be readily accessible to people with disabilities.

2. It directly impacts upon those people who are responsible for the operation and use of facilities, including landlords, tenants, owners, operators and facility managers.

This mandate directly impacts upon facility managers because Title III regulations cover:

- Private entities that own, operate, lease from or lease to places of public accommodation.

- Commercial facilities.

- Private entities that offer certain examinations and courses related to educational and occupational certification.

Like the Title I employment provisions and the Title II provisions which regulate programs, services and activities of state and local government, Title III is a civil rights law that prohibits discrimination against individuals with disabilities. Its purpose is to promote the accommodation of people with disabilities in the delivery and receipt of goods and services.

TITLE I AND THE FACILITY MANAGER

Unlike the standards that regulate state and local government programs, services and activities, or public accommodations and commercial facilities (Title II and Title III, respectively), the employment provisions of ADA's Title I are less "clear cut" and definitive.

For the facility manager who is responsible for integrating the employee with the work environment, this issue is compounded by three different, but interrelated, factors.

First, the preponderance of disability-related discrimination complaints filed with the Equal Employment Opportunity Commission (EEOC), which enforces Title I of the ADA, are work-injury related.

Second, confusion surrounds an injured worker's "status" as an individual with a disability. While the ADA does not specifically identify an injured worker as a person with a disability, it does not exclude an injured worker from its definition of a person with a disability. The key to determining the injured worker's "status" rests with the worker, meeting the ADA's definitional test. If the injured worker does meet ADA criteria, that individual assumes the same rights as other qualified individuals under the ADA.

Third, the impact of workers' compensation upon the injured worker adds to the complexity, especially in terms of benefits entitlement, fitness to return to work with or without reasonable accommodation, and the employer's ability/inability to reasonably accommodate the injured worker with a disability.

The ADA's Definition of Disability

The ADA defines a person with a disability as an individual who:

- Has a physical or mental impairment that substantially limits one or more major life activities.
- Has a "record" of such an impairment.
- Is regarded as having such an impairment.
- Is associated with an individual with a disability.

Physical and Mental Impairments

A physical impairment is defined as any physiological disorder or condition, cosmetic disfigurement or anatomical loss affecting one or more of the following body systems: neurological, respiratory (including speech organs), cardiovascular, reproductive, digestive, genital-urinary, hemic and lymphatic, skin and endocrine.

A mental impairment is defined as any mental or physiological disorder, such as mental retardation, organic brain syndrome, emotional or mental illness, and specific learning disabilities.

The ADA does not provide a comprehensive list of physical and mental impairments. The number and variety of impairments preclude it from doing so. The impairment itself, however, must either be either physiological or mental. Simple physical characteristics such as height/weight within a normal range, left handedness, hair/eye color, etc., are not impairments.

Determination of impairment is made without regard to any kind of medication or device that an individual may be required to take or use. A person required to use a vasodilator to control asthmatic attacks, for example, is considered to have an impairment, even if the use of the vasodilator completely prevents attacks.

Persons who have contracted contagious diseases such as tuberculosis, or Human Immunodeficiency Virus (HIV), are considered to have an impairment under the ADA.

While the ADA recognizes drug addiction as an impairment, employees (and applicants) are not protected from any personnel actions based on current use of illegal drugs. Individuals who are no longer illegally using drugs and who have been successfully rehabilitated, or are in the process of completing a rehabilitation program, however, are protected under the Act.

Stress and depression may be considered impairments under EEOC guidelines. Determination is based upon documented physiological or mental assessments.

Major Life Activities

The impairment must substantially limit one or more major life activities for the individual to be considered as having a disability under the ADA.

Major life activities are defined as those basic activities that the average person in the general population can perform with little or no difficulty. EEOC Guidelines list the following as representative of major life activities:

walking	speaking	breathing
seeing	hearing	sitting
standing	reaching	lifting
reading	manual tasks	working
caring for one's self		

Substantially Limiting

Under the ADA, an impairment is a disability if it "substantially limits" one or more major life activities. There are three factors that are used to determine if an impairment substantially limits a major life activity:

- The nature and severity of the impairment.
- The duration/expected duration of the impairment.
- The long-term or permanent impact/or expected long-term or permanent impact of the impairment.

It is important to consider these, because it is the *effect* of the impairment that determines if one or more of a person's major life activities has been substantially limited. The same holds true for the individual who has multiple impairments.

While no one impairment may substantially limit a major life activity, the cumulative effect of those impairments may limit a major life activity. When and if this occurs, the individual does have a disability.

Has A "Record" of Such an Impairment

Under the ADA, people who have a history of disability are protected, whether or not they are currently substantially limited in a major life activity. People who have either been wrongly classified or misdiagnosed as having a disability are also protected under the ADA.

Is Regarded as Having an Impairment

Perception of impairment is a significant factor. If a person is perceived to have an impairment that substantially limits a major life activity, then that person is protected under the ADA.

Temporary Impairments

The duration of an impairment is not the only factor that determines substantial limitations.

The extent of the impairment, as well as the impact, or effects of the impairment are factors that must be considered. A temporary, non-chronic impairment that does not last for any given length of time, and that does not have a long-term impact upon the individual, is not considered a disability for the most part.

Qualified Individuals with Disabilities

For an individual to be protected by the ADA, the individual must be qualified as such. In other words, a qualified individual with a disability is a person who meets the prerequisite skills, experience, educational background and all other requirements of the job, and who can perform the essential functions of the job, with or without reasonable accommodation.

If the person meets the necessary prerequisites of the job, then the person is considered "otherwise qualified."

In this case, the hiring manager must determine if that person can perform the essential functions of the job with or without reasonable accommodations.

In addition to their own evaluation, or judgment as to whether or not the person can perform the essential functions of the job, the hiring manager must consider:

- The job description (job descriptions must accurately reflect the functions of the job the employee will be performing).
- The actual amount of time the person will spend performing the functions of the job.
- Any consequences of not requiring the person to perform the functions of the job.
- The work experience of people who have performed the job in the past, as well as persons who are currently doing the job.
- The nature of the work operation.
- Collective bargaining agreements (if applicable).

If the individual cannot perform the essential function of the job, with or without reasonable accommodation then that person is not considered qualified by the ADA.

Reasonable Accommodation

The ADA requires that employers reasonably accommodate current employees and job applicants with disabilities. The general rule states than an employer cannot select a qualified person without a disability over an equally qualified person with a disability only because the employee/applicant with a disability will require accommodation.

Some examples of accommodation include making existing facilities accessible to people with disabilities, job restructuring, equipment and/or worksite modification, provision of assistance aids (e.g., Braille materials, etc.) and job reassignment if the person can no longer perform his original job duties.

Because reasonable accommodation must be made on an individual basis, the EEOC holds that reasonable accommodation must be jointly determined by employer and employee (or applicant). This is accomplished by consulting with the qualified employee or applicant with a disability to identify limitations, as well as appropriate agencies, rehabilitation organizations and medical care providers to determine potential accommodations. The accommodation appropriate for the person to perform the essential functions of the job is then chosen.

Note that while the accommodation must be reasonable, it does not need to be the best. If a less expensive assistance aid (e.g., manually-operated wheelchair) will permit the person to perform the essential job functions, for example, the employer does not need to provide a better or more expensive aid (e.g., motorized wheelchair). The employee (or applicant) may, however, choose a different accommodation if he is willing to provide it.

TITLE II AND THE FACILITY MANAGER

The significance of Title II rests with the provision of programs, services and activities by state and local government entities.

With the exception of public transportation services that are operated by state and local governments, any public entity—including departments, agencies, or other divisions of state and local government—are subject to Title II provisions.

(Note: The United States Department of Transportation's (U.S. DOT's) regulations establish specific requirements for transportation vehicles and facilities, including a requirement that all new busses be equipped to provide services to people who use wheelchairs.)

Compliance with Title II

State and local governments are required to:

- Permit any and/or all persons with disabilities to participate in any and/or all programs, services or activities under the entity's jurisdiction.

- Provide programs and services in an integrated setting, unless separate or different measures are necessary to ensure equal opportunity.

- Eliminate unnecessary eligibility standards or rules that deny individuals with disabilities equal opportunity to enjoy the programs, services or activities under the jurisdiction of the public entity, unless doing so is "necessary" for the ongoing provision of the program, service or activity; as well as any requirement that ends to screen out individuals with disabilities is prohibited; and any safety program, however, such as eligibility for drivers' licenses, may be imposed if they are based on actual risks and not on mere speculation, stereotypes or generalizations about individuals with disabilities.

- Furnish auxiliary aids and services, when necessary, to ensure effective communication, unless an undue burden or fundamental alteration would result.

- Make reasonable modification to policies, practices and procedures that deny equal access to individuals with disabilities, unless a fundamental alteration to the program would result.

- Operate their programs so that, when viewed in their entirety, the programs are readily accessible to and usable by individuals with disabilities.

Public entities are also prohibited from placing special charges on individuals with disabilities to cover the costs of measures necessary to ensure nondiscriminatory treatment, such as making modifications required to provide program accessibility or providing qualified interpreters.

Generally speaking, however, the areas that have the greatest impact upon facility managers are program access, including integrated programs, communications and new construction and alterations. Each of these areas, however, is addressed in terms of the eligibility requirements of the individual with a disability.

Eligibility Criteria

To repeat, an individual with a disability is a person who has a physical or mental impairment that substantially limits a "major life activity"; has a "record" of such an impairment; is regarded as having such an impairment; or is associated with an individual with a disability.

The "qualified" individual with a disability is a person who meets the essential eligibility requirements for the program, service or activity offered by the public entity.

"Essential eligibility requirements" depend on the type of program service or activity that is involved. In some activities, such as a state licensing program, the ability to meet specific skill and performance requirements may be "essential." The "essential eligibility requirements" would be minimal in places where the public entity provides information to anyone requesting it.

Also note that although the ADA recognizes drug addiction as an impairment, state and local government entities may withhold any services or benefits to an individual who currently engages in the use, possession or distribution of illegal drugs. In making a determination to withhold services and benefits, the public entity must carefully review all facts to ensure that its belief is reasonable. Individuals who have completed a supervised drug rehabilitation program, or who are currently participating in a supervised drug rehabilitation program and who are not otherwise engaged in illegal drug activity, are protected under Title II of the ADA.

Program Access

Under Title II provisions, state and local governments are required to:

- Ensure that persons with disabilities are not excluded from programs, services and activities because buildings are inaccessible.

- Make the programs accessible to individuals who are unable to use an existing facility that is inaccessible.

- Provide alternative delivery methods for the programs, services and activities currently offered in an inaccessible building (e.g., relocating a service to the first floor of a building; providing personal assistance to enable a person with a disability to obtain the service; providing benefits or services at the individual's home or at an alternative accessible site).

State and local government entities are not required to take any action that would result in a fundamental alteration in the nature of their programs, services or activities. Nor are they required to take any action that would result in undue financial and/or administrative burdens. However, public entities are required to take any available action that would ensure that persons with disabilities would receive benefits or services, provided that the action would not result in fundamental changes or create undue burden.

Integrated Programs

Integrating persons with disabilities into the mainstream of society is fundamental to the ADA. For this reason, public entities are generally prohibited from offering programs and services that are separate or different from programs and services offered to people without disabilities. Public entities may offer separate programming or services in those instances where separate programming services are necessary to ensure that benefits or services are equally effective, however.

In those instances where a separate program is permitted, the person with a disability still maintains the right to participate in a regular program.

State and local governments may not require any person with a disability to accept an accommodation or benefit. The person with a disability retains the right to refuse accommodation or benefit.

Communications

State and local governments are required to ensure effective communications with persons who have disabilities. To accomplish this, the public entity must provide appropriate auxiliary or assistance aids.

Examples of auxiliary aids, which must be provided at no cost to the person with a disability, include (but are not limited to):

Qualified interpreters	Readers
Assistive listening devices	Taped text
Televised captioning and decoders	Braille materials
Large print materials	TDDs
Videotext displays	

New Construction and Alterations

State and local governments must ensure that newly constructed buildings and facilities are free of architectural and communication barriers that restrict access or use by persons with disabilities.

While the ADA does not require retrofitting of existing buildings to eliminate barriers, it does require the entity to ensure accessibility at the time any alteration is undertaken.

Technical Standards for Accessible Design

Since its enactment, the ADA has provided public entities with the option of choosing between two technical standards for accessible design: The Uniform Federal Accessibility Standard (UFAS), established under the Architectural Barriers Act; and the Americans with Disability Act Accessibility Guidelines (ADAAG), adopted by the U.S. Department of Justice (DOJ) for places of public accommodation and commercial facilities covered by Title III of the ADA.

Enforcement

State and local government entities are subject to the Title II provisions of the ADA, which are enforced by the U.S. Department of Justice. Individuals, as well as classes of individuals who believe they have been discriminated against by any state or local government entity may file complaints with any federal agency that provides financial assistance to the state or local program in question or with the U.S. Department of Justice, which will refer the complaint to the appropriate agency.

Remedies available are the same as provided under section 504 of the Rehabilitation Act of 1973. Reasonable attorneys' fees may also be awarded to the prevailing party.

TITLE III AND THE FACILITY MANAGER

Title III of the ADA mandates that public accommodations and commercial facilities be readily accessible to persons with disabilities. This mandate directly impacts many facility managers because Title III regulations cover:

- Private entities that own, operate, lease from or lease to places of public accommodation.
- Commercial facilities.
- Private entities that offer certain examinations and courses related to educational and occupational certification.

Like Title I and Title II provisions of the ADA, Title III is a civil rights law that prohibits discrimination against people with disabilities. Its purpose is to promote the accommodation of persons with disabilities in the delivery and receipt of goods and services.

Individuals with Disabilities

Again, an individual is considered as having a disability if he has a physical or mental impairment that substantially limits a "major life activity," has a "record" of such an impairment and is regarded as having an impairment, whether he has one or not.

Also included in the definition of impairment is drug addition. In most cases, however, a public accommodation may make a decision to withhold services or benefits to an individual who currently engages in the use, possession or distribution of illegal drugs.

"Current use" is the illegal use of controlled substances that occurred recently enough to justify a reasonable belief that a person's drug use is current, or that continuing use is a real and ongoing problem. In basing a decision to withhold services or benefits from a person engaged in illegal drug activity, the private entity must carefully review all of the available facts to ensure that its belief is reasonable.

People who have successfully completed a drug rehabilitation program, or are currently participating in a supervised drug rehabilitation program and are not currently engaging in illegal drug use, are protected by Title III of the ADA.

Defining Public Accommodations and Commercial Facilities

Public Accommodations

A facility is considered a place of public accommodation when it is operated by a private entity, its operation affects commerce and it falls within one of the categories determined by the ADA to be a place of public accommodation:

- Places of lodging such as hotels and motels, except those owner-occupied places that rent fewer than six rooms.
- Restaurants, bars and other eating and drinking establishments.
- Movie theaters, stadiums, concert halls and other places of entertainment.
- Auditoriums, convention centers, lecture halls and other facilities where people gather.
- Retail and service facilities, such as banks, lawyers and accountant offices, hospitals, offices of health care providers, barber and beauty shops, pharmacies and insurance offices.
- Public transportation terminals, depots and stations with the exception of facilities related to air transportation.
- Libraries, galleries, museums and other places of public display or collection.
- Recreational areas such as amusement parks, zoos and parks.
- Private educational facilities, including graduate and undergraduate schools, secondary, elementary and private nursery schools.
- Social service facilities including senior citizen centers, child day care and homeless centers, food banks and adoption agencies.
- Exercise facilities including golf courses, health spas, bowling alleys and gymnasiums.

In short, a public accommodation is any place that involves the general public and the general public's right to buy, sell or engage in a variety of activities within the facility.

Facilities Not Considered Public Accommodations—There are several places of public gathering that are not considered to be places of public accommodation.

These include any residential apartment/condominium complexes which are subject to Fair Housing Act provisions, religious entities and private clubs. (Religious entities may be subject to the Title I provision of the ADA if their congregations employ the number of persons required for coverage. Private clubs are exempt only if they meet the specific criteria defined in Title II of the Civil Rights Act of 1964).

Defining Commercial Facilities

Commercial facilities, as defined by the ADA, are facilities that are intended for non-residential use by a private entity, and whose operations affect commerce.

Commercial facilities include warehouses, factories, office buildings, airports and wholesale facilities that sell exclusively to businesses.

Facilities Not Considered Commercial Facilities—Generally speaking, a commercial facility involves only a select group of individuals. For example, only a company's employees and business associates are authorized to engage in activities within the facility. Any facility that is covered by the Fair Housing Act of 1968, such as residential apartment complexes, are specifically exempted from the ADA's definition of commercial facilities.

Combined Facilities

Companies that operate as a "mixed-use" or combined facility are subject to Title III provisions as well. A company that operates a warehouse that sells goods to a specific trade, for example, is not considered a public accommodation. However, those companies which sell their goods/products to the general public in any company-operated outlet store or franchise must comply with the public accommodations provision of the ADA.

Public Accommodations: Employer Responsibilities

Public accommodations under the ADA carry specific obligations for employers. These obligations include:

- Removing architectural barriers and communications barriers that are structural in nature, to the extent that it is "readily achievable." Cost and the necessity for making the change(s) determine if removal of barriers is readily achievable.

- Determine if other alternatives are available when barrier removal is not readily achievable.

- Provide for auxiliary aids and/or services to persons with disabilities when and where necessary to prevent those persons from being excluded, denied service, segregated or otherwise treated differently than the general population.

- Make reasonable modifications to policies, practices and procedures that are necessary to providing equal goods and services to persons with disabilities.

- Review any eligibility requirements or admissions criteria to ensure that persons with disabilities are not excluded.

Removal of Architectural Barriers

A facility that meets the ADA's definition of public accommodation must remove architectural barriers when it is readily achievable. While there is no requirement to remove a barrier from an area that is accessible only to employees, the area itself may fall under the regulatory scope of Title I employment provisions.

Priorities for Barrier Removal—The U.S. Department of Justice, which enforces Title III provisions of the ADA, has determined priorities for removal of barriers. The purpose of these priorities is to facilitate long-term business planning and to maximize the degree of effective access that will result from any given level of expenditure.

The priorities are not mandatory. Public accommodations have the latitude to determine the most effective "mix" of barrier removal measures to undertake in their facilities.

The first priority is to enable persons with disabilities to physically enter the facility. This priority is generally preferable to other alternative arrangements in terms of both business efficiency and the dignity of persons with disabilities.

The second priority is for measures that provide access to those areas of a place of public accommodation where goods and services are made available to the general public. To the extent that it is readily achievable to do so, persons with disabilities should be given access to front desk assistance in a store, as well as access to the retail display areas of the store, for example.

The third priority should be providing access to restrooms, if restrooms are provided for use by customers and clients.

The fourth priority is to remove any barriers to using a place of public accommodations facility (for example, lowering public telephones).

Readily Achievable—For an architectural barrier to be "readily achievable," it must be "easily accomplishable and able to be carried out without much difficulty or expense." Factors that must be considered in removing a barrier are the nature of the barrier and its cost, the financial resources of the site involved, the distance of the site from the parent company or other entity, the financial resources of these parties, and the type of business operation of these parties.

If the barrier removal is not readily achievable, the facility must make its goods and services available to persons with disabilities through alternative methods. Any measures taken to provide alternative methods, however, cannot pose a significant risk to the safety or health of persons with disabilities, or others.

When a facility determines that installing a permanent ramp to accommodate persons who use wheelchairs is not readily achievable, for example, a portable ramp may be provided. To ensure safety, the portable ramp should have railings and a firm and stable non-slip surface. The portable ramp should also be properly secured.

Provide Auxiliary Aids

Facilities that meet the definition of public accommodations must provide auxiliary aids to ensure equal accessibility unless an undue burden or fundamental alteration would result.

Various types of auxiliary aids that must be provided include:

Qualified interpreters

Telecommunications devices for deaf persons (TTDs)

Closed captioned decoders

Qualified readers for sight-impaired persons

Braille materials

Audio tapes

TTDs for persons with speech impairments

Computer terminals

Speech synthesizers

Undue Burden—Public accommodations are not required to provide auxiliary aids if doing so would result in an undue burden. The same criteria used to determine the feasibility of removing barriers apply to making an undue burden determination.

Fundamental Alteration—Public accommodations are not required to provide auxiliary aids for people with disabilities, if a fundamental alteration in the nature for the goods and/or services would result.

Modify Policies To Avoid Discrimination

Facilities that are public accommodations must reasonably modify policies, practices and procedures to avoid discriminating against people with disabilities. Sight-impaired persons who use service animals (i.e., guide dogs), for example, must be permitted entry into a public accommodation, unless doing so would result in either a fundamental alteration, or if the safe operation of the facility would be jeopardized.

Public accommodations are prohibited from charging a fee to persons with disabilities to cover the cost of the accommodation. For example, a fee for home delivery of goods and/or services may not be charged to persons with disabilities, unless the general population is charged the same fee for home delivery.

Ensure Non-discriminatory Eligibility Criteria

Public accommodations are prohibited from establishing eligibility criteria that tend to screen out persons with disabilities. A violation of Title III provisions would result, for example, if a bank required a valid driver's license as proof of identity for new depositors.

Public accommodations are also prohibited from making unnecessary inquiries into the existence of a disability. A private summer camp, for example, may require parents to complete a questionnaire and to submit medical documentation regarding their child's ability to participate in the various activities. The questionnaire, itself, is acceptable if the camp can demonstrate that each piece of information requested is needed to ensure the child's safe participation in activities. The camp is prohibited from using the medical information to screen out children from the camp, however.

Places of public accommodation may impose any legitimate safety requirements that are necessary for safe operation. A wilderness tour company may require participants to meet a certain level of swimming proficiency to participate in a rafting expedition, for example.

Constructing and Renovating Commercial Facilities—Accessibility Guidelines

In July 2004, the U.S. Access Board, an independent Federal agency, issued updated accessibility guidelines for new or altered facilities covered by the Americans with Disabilities Act and the Architectural Barriers Act. These guidelines address a wide range of facilities in the private and public sectors. Determining which standard applies depends, in large part, on the type of facility. For example, ADA standards govern private sector facilities and state and local government facilities. ABA standards, however, govern federally funded facilities.

BACKGROUND REVIEW

The Americans With Disabilities Act (ADA) of 1990

The ADA, a major civil rights law prohibiting discrimination on the basis of disability, establishes design requirements for the construction or alteration of facilities. It covers facilities in the private sector—places of public accommodation and commercial facilities—and the public sector—state and local government facilities. Under the ADA, the Board is responsible for accessibility guidelines covering newly built and altered facilities. In 1991, the Board published the ADA Accessibility Guidelines (ADAAG) which serve as the basis for standards used to enforce the law. The new guidelines overhaul the original ADAAG.

The Architectural Barriers Act (ABA) of 1968

The ABA requires access to facilities designed, built, altered, or leased with Federal funds. Similar to its responsibility under the ADA, the Board maintains guidelines under the ABA which serve as the basis for the enforceable standards. The Board has updated its guidelines for

ABA facilities jointly with the new ADA guidelines so that a consistent level of access is specified under both laws.

Development of Guidelines

Guidelines are developed and updated using a process common to most Federal regulations. This process provides an opportunity for public comment. To get input from a cross section of stakeholders the Board established an advisory committee. The function of the advisory committee was to review the original guidelines and to recommend changes. The ADAAG Review Advisory Committee, which consisted of 22 members representing the design and construction industry, the building code community, and people with disabilities, as well as other interested parties, submitted a report to the Board that detailed recommended revisions to the substance, organization, and format of the guidelines. The finalized guidelines are based largely on these recommendations. The Board published the guidelines in proposed form in November, 1999 and made them available for public comment for six months. During this public comment period, over 2,500 public comments were presented on the proposed guidelines. Finalized guidelines reflect these comments.

Updates

Key updates include:

- updating specifications so that they continue to meet the needs of persons with disabilities;
- improving the format and usability of the guidelines to facilitate compliance;
- blending the guidelines with model building codes and industry standards; and
- making the requirements for ADA and ABA facilities consistent.

Model Building Codes and Industry Standards

In order to make compliance easier, efforts were made to make the guidelines more consistent with model building codes and industry standards. Model code groups and standard-setting bodies were coordinated extensively so that differences could be reconciled. In particular, the Board sought to harmonize, or blend, the guidelines with the International Building Code (IBC) and access standards issued through the American National Standards Institute (ANSI). Used by a growing number of states and local jurisdictions, the IBC contains scoping provisions for accessibility. The ANSI A117.1 standard, a voluntary consensus standard, provides technical criteria referenced by the IBC. A number of revisions were made to the guidelines for consistency with these and other model codes and standards. In addition, the Board worked to resolve remaining differences by advocating changes to the IBC and the ANSI A117.1 standard based on the new guidelines.

Timeframes

The Board's guidelines are not mandatory on the public. However, the guidelines serve as the baseline for enforceable standards which are maintained by other Federal agencies. The guidelines are similar to model building codes, in that they are not required to be followed except in those cases when the guidelines are adopted by an enforcing authority. Under the ADA, the Department of Justice, along with the Department of Transportation, with respect to transit facilities are responsible for enforceable standards based on the Board's guidelines. These agencies will update their ADA standards based on the new guidelines and indicate when the new standards are to be followed. Several other agencies including the General Services Administration, the Department of Defense, the Department of Housing and Urban Development, and the U.S. Postal Service also hold a similar responsibility for enforcing ABA standards.

Existing Facilities

The ADA and ABA guidelines cover new construction and planned alterations. These guidelines generally do not apply to existing facilities except where altered. Facilities built or altered according to earlier versions of the ADA or ABA standards will not necessarily have to meet the updated version except when they are subsequently altered or renovated. In order to update the existing standards, the Department of Justice, which regulates requirements for existing facilities under the ADA, will address coverage of facilities built or altered according to the original ADA standards. Also addressed are facilities retrofitted under ADA provisions for existing facilities. For example, the requirement for barrier removal in places of public accommodation will be outlined.

With respect to ABA facilities, the Access Board has clarified that facilities built to earlier ABA standards are subject to the new requirements only in relation to planned alterations.

OVERVIEW

The updated guidelines feature:

- a new numbering system consistent with model codes;
- a more streamlined structure and organization of chapters;
- updated scoping and technical provisions, with a greater structural delineation between them;
- new figures and commentary—advisory information; and
- provision of all figure-based information in written text.

A Rule in Three Parts

The Board coordinated its update of the ADA and ABA guidelines into a single rule. The final rule contains updated scoping provisions, which specify accessibility requirements as well as the technical requirements needed for achievable accessibility. The rule contains three parts: a scoping document for ADA facilities, a scoping document for ABA facilities, and a common set of technical criteria referenced by both scoping documents.

Supplements to ADAAG

The U.S. Access Board previously developed supplements to the original ADA guidelines that are specific to different types of facilities and elements:

- state and local government facilities, including courthouses and prisons;
- building elements designed for children's use;
- play areas; and
- recreation facilities.

These supplements are included in the new guidelines. They have been revised for consistency with the format and approach, but their substance remains unchanged.

SUMMARY

ADA Application and Scoping

The guidelines include general provisions that recognize the purpose of the guidelines, specifications for adults and children, equivalent facilitation, which permits departures providing equal or greater access, conventions, referenced standards, and definitions. These provisions include instructions on applying the guidelines, such as conventions concerning specified dimensions. Throughout the guidelines, the Access Board has replaced absolute dimensions with specified ranges wherever practicable to facilitate compliance.

The guidelines reference several model building codes and industry standards. These include industry standards for powered doors, elevators, platform lifts, and play surfacing and equipment. Requirements for means of egress and fire alarms are addressed through references to the International Building Code (IBC) and the National Fire Alarm Code (NFPA 72). The guidelines reference the most recent editions of these codes and standards.

Scoping Requirements

The format and content of the guidelines reinforce the underlying premise that all areas of newly constructed facilities are required to be accessible unless otherwise noted. Consequently, exceptions from the requirements are more thoroughly covered.

The new guidelines enhance coverage of employee work areas. The original ADA guidelines specified that work areas be on an accessible route so that persons with disabilities can approach, enter, and exit the space. In addition to this, the new guidelines also require the accessibility of circulation paths within sizable work areas (1000 square feet or more). They also address accessible means of egress from work areas and connections for visual alarms. Another notable revision concerns press boxes, which by their elevation and location have posed challenges to access. The new guidelines include an exception for certain press boxes based on their size, elevation, and location.

Other revisions include:

- enhanced scoping for public entrances, van parking, passenger loading zones, stairways, and telecommunication devices (TTYs) at pay phones for persons with hearing or speech impairments;

- new or clarifying provisions covering access to different types of elevators including destination-oriented, limited-use/limited application or "LULA," and residential elevators. Provisions also include access to drinking fountains, kitchens, kitchenettes, and sinks, washing machines and clothes dryers, signs, dispersed wheelchair seating, windows, and residential dwelling units; and

- reduced scoping for unisex toilet rooms located at a single location—half instead of all, and for wheelchair spaces in large assembly areas.

ABA Application and Scoping

Application and scoping requirements for ABA facilities are based on the application and scoping requirements for ADA facilities to ensure a consistent level of access. There are differences in certain areas which stem from variations between the ADA and ABA statutes. For example, the ABA is broader in its coverage of employee work areas, a difference reflected in the updated guidelines. Exceptions for work areas that limit coverage in the ADA scoping document are not included in the ABA counterpart. Other ABA provisions that differ from the ADA document concern modifications and waivers, definitions, additions, leased facilities, existing elements, and residential facilities.

Basic Elements: The Building Blocks of Accessibility

The basic elements considered to be the "building blocks" of accessibility as established by the guidelines include ground and floor surfaces, changes in level, wheelchair turning space, clear floor space, knee and toe clearances, protruding objects, reach ranges, and operable parts.

The guidelines specify reach ranges according to the approach either forward or side. A significant change reduces the maximum side reach range from 54 to 48 inches, the height specified for forward reaches. This change, which was recommended by the ADAAG Review Advisory Committee and strongly supported by public comments, includes exceptions for certain elements, such as gas pumps.

Accessible Routes

New specifications for accessible routes that include walking surfaces, doors, ramps, curb ramps, elevators, and platform lifts are provided. These specifications clarify access at recessed doors, ramps (edge protection), and curb ramps (top landings). Provisions for elevators recognize a greater range of designs and dimensions for standard cars and include new technical criteria for other types of elevators.

These include destination-oriented, limited-use/limited-application, and residential elevators. Provisions for platform lifts have been updated and reference a new industry standard (ASME A18.1).

The original guidelines required detectable warnings, a distinctive tactile surfacing, on the surface of curb ramps and other areas. This requirement alerts persons with vision impairments of their approach to streets and drop-offs at boarding platforms. The new guidelines do not include a requirement for detectable warnings at curb ramps or hazardous vehicular areas; this issue is being revisited in a separate rulemaking on accessible public rights-of-way. Under the new guidelines, detectable warnings are still required along the edges of boarding platforms in transit facilities.

General Site and Building Elements

These guidelines contain requirements for parking, passenger loading zones, stairways, and handrails. Revisions of the guidelines include a new provision specific to angled van parking spaces and revamped specifications for handrails that will permit a greater range of designs and shapes.

Plumbing Elements and Facilities

The specifications for plumbed fixtures address drinking fountains, toilets and bathrooms, water closets and compartments, urinals, lavatories and sinks, bathtubs, showers, grab bars, tub and shower seats, washing machines and clothes dryers, and saunas and steam rooms.

Some provisions have been revised to help improve compliance as well as access. For example, an absolute dimension for the centerline placement of toilets—formerly 18 inches—has been replaced with a range of 16 to 18 inches. To improve access and allow side transfers at toilets, lavatories are no longer permitted to overlap the required clear space aside toilets.

Other changes include drinking fountains (side approach access is no longer permitted at wheelchair accessible units), shower compartments and specifications for water temperature, as well as spray units, have also been changed.

Communication Elements and Features

Technical criteria for communication elements such as fire alarms, signs, telephones, detectable warnings, assistive listening systems, ATMs and fare machines, and two-way communication systems is provided. Substantive changes include:

- addressing technical criteria for fire alarms through the National Fire Alarm Code (NFPA 72), which effectively overhauls specifications for visual alarms in a manner that facilitates compliance while enhancing design and installation options;
- revamped specifications for signs;
- new specifications for the capabilities and sound quality of assistive listening systems that derive from Board-sponsored research;
- improved access at ATMs and fare machines for persons with vision impairments through detailed criteria for audible output and tactile markings; and
- revised specifications for detectable warnings to allow a greater range of designs and products.

Special Rooms, Spaces, and Elements

Various types of occupancies and spaces are addressed including: assembly areas, dressing, fitting, and locker rooms, kitchens and kitchenettes, medical care facilities, transient lodging, holding and housing cells, courtrooms, residential dwelling units, transportation facilities, and storage.

The new guidelines are structured to make provisions for certain types of facilities or spaces consistent. Some provisions specific to these facilities also cover elements addressed for other facilities. For example, provisions specific to toilet and bathing facilities in residential dwelling units are also noted.

The additional requirements were reorganized to clarify the application of requirements for certain types of spaces without respect to the overall occupancy. For example, specifications for kitchens and kitchenettes apply whether such spaces are located in a hotel guest room, a dwelling unit, or an employee break room. This differs from the original guidelines which addressed kitchens and kitchenettes only in relation to transient lodging facilities and, in the case of the ABA guidelines, dwelling units.

Built-in Furnishings and Equipment

Specifications covering built in equipment and furnishings as well as dining and work surfaces, benches, and sales and service counters, including check-out aisles are also noted. These guidelines provide revised specifications for benches which include revised criteria for back support.

Recreation Facilities and Play Areas

Technical provisions for various types of recreation facilities, including play areas developed previously as supplements to the original ADA guidelines are also listed. They have been integrated into the new guidelines without substantive change. Requirements are provided for amusement rides, recreational boating facilities, exercise machines, fishing piers and platforms, golf facilities, miniature golf facilities, play areas, swimming pools, wading pools, and spas, and shooting facilities with firing positions.

ADA and ABA Guidelines: Supplementary Information

On March 23, 2007 supplementary information on the Americans with Disabilities Act (ADA) and the Architectural Barriers Act (ABA) was added. This information, which does not affect the substance of the guidelines, provides information on the adoption of enforceable standards based on these guidelines under the ADA and ABA. This information contains details, including effective dates, for new ADA standards adopted by the U.S. Department of Transportation and updated ABA standards implemented by the General Services Administration and the U.S. Postal Service.

Sources

Gustin, Joseph F. *Safety Management: A Guide For Facility Managers*, New York: UpWord Publishing, Inc., 1996.

United States Department of Justice, *Code of Federal Regulations*, 28 CFR Part 36, Final Rule.

United States Department of Justice, Civil Rights Division Disability Rights Section, *Title II Highlights*.

United States Department of Justice, Civil Rights Division Disability Rights Section, *Title III Highlights*.

United States Department of Labor, Occupational Safety And Health Administration, "Hazardous Waste Operations And Emergency Response," Title 29, *Code of Federal Regulations*, 1910.120.

United States Department of Labor, Occupational Safety And Health Administration,

Principal Emergency Response And Preparedness Requirements in OSHA Standards And Guidance for Safety And Health Programs, OSHA 3122, 1993 (Rev.).

United States Department of Labor, Occupational Safety And Health Administration, "Process Safety Management of Highly Hazardous Chemicals," Title 29, *Code of Federal Regulations, Part 1910.119.*

United States Access Board, *Supplementary Information to the ADA and ABA Guidelines* March, 2007.

Chapter 6

Assessing and Managing Risk

The terrorist attacks of September 11, 2001, clearly demonstrated that our institutions and our infrastructure have been targeted. The terrorist's weapons of mass destruction can include biological, chemical, radiological, nuclear, or high explosive weaponry. And, while terrorist groups would have to overcome significant technical and operational challenges to make and release many chemical and biological agents of a sufficient quality and quantity to kill large numbers of people, it has been tried.

Recent examples of terrorist activity, both at home and abroad, include the incidents of anthrax-laced letters sent to public and high-profile citizens following the September 11th attacks. In the release of the nerve agent sarin in a Tokyo subway in 1995, 12 people were killed and thousands were injured. In 1984, a religious cult in Oregon contaminated salad bars in local restaurants with salmonella bacteria in order to prevent people from voting in a local election. Although there were no fatalities as a result of this terrorist attack, hundreds of people were diagnosed with food-borne illness.

As both private and public sector organizations become more aware of their vulnerability to terrorist attacks, they are also becoming acutely aware of the need to increase security measures. In order to better prepare against terrorism and other threats, facility managers and building owners are reviewing their policies and procedures with an eye toward risk management.

RISK MANAGEMENT

Risk management is the systematic, analytical process that considers the likelihood of a threat harming individuals or physical assets. Risk management identifies actions that reduce the risk and mitigate the consequences of an attack or event. Risk management acknowledges that risk generally cannot be eliminated but risk can be reduced by enhancing protection from known or potential threats. These risk management principles include:

- Identify weaknesses in a company, system or organization;
- Offer a realistic method for making decisions about the expenditure of scarce resources and the selection of cost-effective countermeasures to protect assets;
- Improve the success rate of a company/organization's security efforts by emphasizing the communication of risks and recommendations to the final decision-making authority; and
- Assist facility managers, building owners and security professionals as well as other key decision makers answer the question "How much security is enough?"

Since risk is a function of assets, threats and vulnerabilities, risk management allows organizations and companies to determine the:

- magnitude and effect of potential loss;
- likelihood of such loss actually happening; and finally,
- countermeasures that could lower the probability or magnitude of the loss.

As defined by the National Infrastructure Protection Center (NIPC), risk is the potential for some unwanted event to occur. Examples of such unwanted events include loss of information and money, as well as organizational reputations, or someone gaining unauthorized access to the company's physical property, data systems, etc. As such, risk is the function of the likelihood of the unwanted event occurring and its consequences. Therefore, it is obvious that the higher the probability and the greater the consequences, the greater the risk to the company or organization.

The likelihood of the unwanted event occurring depends upon threat and vulnerability. Threat is the capability and the intention of a terrorist to undertake actions that are detrimental to a company or organization's interests. Threat is a function of the terrorist only. It cannot be controlled by the owner or the user of the asset, i.e., building owner, facility/property manager, etc. However, the intention of terrorists to exploit their capability may be encouraged by vulnerability in the company or organization's assets. Or conversely, the intention of the terrorist may be discouraged by the owner's countermeasures.

Vulnerability is any weakness in an asset that can be exploited by a terrorist to cause damage to the company/organization's interest. The level of vulnerability and level of risk can be reduced by implementing appropriate security countermeasures.

An asset is anything of value. Assets include:

- people
- information
- hardware
- software
- facilities
- company reputation
- company operations

In other words, assets are what a company or organization needs to get the job done. Consequently, the more critical the asset is to a company or organization, and the more critical the asset is to meeting company goals, the greater the effect of damage and/or destruction to the asset.

Risk Assessment

The first step in the process of assessing risk to a terrorist attack is to identify the relative importance of the people, business activities, goods and facilities involved in order to prioritize security actions. This applies to both new and existing facilities. The Federal Emergency Management Agency (FEMA) recommends the following actions:

- Define and understand the core functions and process of the business or institutional entity

- Identify critical business infrastructure:
  - Critical component—people, functions and facilities
  - Critical information systems and data
  - Life safety systems and safe haven areas
  - Security systems

- Assign a relative protection priority such as high, medium or low to the occupants, business functions or physical components of the facility:
  - High Priority—loss or damage of the facility would have grave consequences such as loss of life, severe injuries, loss of primary services or major loss of core processes and functions for an extended period of time.
  - Medium Priority—loss or damage of the facility would have moderate to serious consequences such as injuries or impairment of core functions and processes
  - Low Priority—loss or damage of the facility would have minor consequences or impact, such as a slight impact on core functions and processes for a short period of time.

Threat Assessment

Terrorist threat comes from people with the intent to do harm, who are known to exist, have the capability for hostile action, and have expressed the intent to take hostile action.

Threat assessment is a continual process of compiling and examining information concerning potential threats. Information should be gathered from all reliable sources. The assessment process consists of:

- Defining threats and
- Identifying likely threat event profiles and tactics.

Defining Threats

Defining threats involves analyzing the following information regarding terrorists:

- Existence
- Capability
- History
- Intention
- Targeting

Existence is the assessment of who is hostile to the organization.

Capability is the assessment of what weapons have been used in carrying out past attacks.

History is the assessment of what the potential terrorist has done in the past and how many times.

Intention is the assessment of what the potential terrorist hopes to achieve.

Targeting is the assessment of the likelihood that a terrorist performing surveillance on a particular facility, nearby facilities, or facilities that have much in common with a particular organization.

Identifying Likely Threat Event Profiles and Tactics

To identify the likelihood of specific threats and tactics, the following variables should be evaluated:

- attack intentions;
- hazard event profiles; and
- the expected effects of an attack on the facility/organization.

Table 6-1 presents general event profiles for a range of possible forms of terrorist attacks. The profiles describe the mode, duration and extent of the effects of an attack, as well as any mitigating and exacerbating conditions that may exist. These descriptions can be used to identify threats of concern to individual organizations.

Assigning a Threat Rating

A threat rating should be assigned to each hazard of concern to a particular organization. The threat rating, like protection priority, is based on expert judgment. For purposes of simplicity the ratings may read high, medium or low:

Table 6-1. Event Profiles for Terrorism and Technological Hazards

Hazard/Threat	Application Mode	Hazard Duration	Extent of Effects; Static/Dynamic	Mitigating and Exacerbating Conditions
Agriterrorism	Direct, generally covert contamination of food supplies or introduction of pests and/or disease agents to crops and livestock.	Days to months.	Varies by type of incident. Food contamination events may be limited to discrete distribution sites, whereas pests and diseases may spread widely. Generally no effects on built environment.	Inadequate security can facilitate adulteration of food and introduction of pests and disease agents to crops and livestock.
Armed Attack - Ballistics (small arms) - Stand-off weapons (rocket propelled grenades, mortars)	Tactical assault or sniping from remote location.	Generally minutes to days.	Varies, based upon the perpetrators' intent and capabilities.	Inadequate security can allow easy access to target, easy concealment of weapons, and undetected initiation of an attack.

Source: FEMA

Table 6-1. Event Profiles for Terrorism and Technological Hazards (*Continued*)

Hazard/Threat	Application Mode	Hazard Duration	Extent of Effects; Static/Dynamic	Mitigating and Exacerbating Conditions
Arson/Incendiary Attack	Initiation of fire or explosion on or near target via direct contact or remotely via projectile.	Generally minutes to hours.	Extent of damage is determined by type and quantity of device /accelerant and materials present at or near target. Effects generally static other than cascading consequences, incremental structural failure, etc.	Mitigation factors include built-in fire detection and protection systems and fire-resistive construction techniques. Inadequate security can allow easy access to target, easy concealment of an incendiary device and undetected initiation of a fire. Non-compliance with fire and building codes as well as failure to maintain existing fire protection systems can substantially increase the effectiveness of a fire weapon.
Biological Agents - Anthrax - Botulism - Brucellosis - Plague - Smallpox - Tularemia - Viral hemorrhagic fevers - Toxins (Botulinum, Ricin, Staphylococcal Enterotoxin B, T-2 Mycotoxins)	Liquid or solid contaminants can be dispersed using sprayers/aerosol generators or by point or line sources such as munitions, covert deposits, and moving sprayers.	Biological agents may pose viable threats for hours to years, depending on the agent and the conditions in which it exists.	Depending on the agent used and the effectiveness with which it is deployed, contamination can be spread via wind and water. Infection can be spread via human or animal vectors.	Altitude of release above ground can affect dispersion; sunlight is destructive to many bacteria and viruses; light to moderate winds will disperse agents but higher winds can break up aerosol clouds; the micro-meteorological effects of buildings and terrain can influence aerosolization and travel of agents.

Source: FEMA

- **High Threat**. Known terrorists or hazards, capable of causing loss and/or damage to a facility exist. One or more vulnerabilities are present and the terrorists are known or are reasonably suspected of having intent to attack the facility.

- **Medium Threat**. Known terrorists or hazards that may be capable of causing loss of or damage to a facility exists. One or more vulnerabilities may be present. However, the terrorists are not believed to have intent to attack the facility.

- **Low Threat**. Few or no terrorists or hazards exist. Their capability of causing damage to a particular facility is doubtful.

Table 6-1. Event Profiles for Terrorism and Technological Hazards (*Continued*)

Hazard/Threat	Application Mode	Hazard Duration	Extent of Effects; Static/Dynamic	Mitigating and Exacerbating Conditions
Chemical Agents - Blister - Blood - Choking/lung/pulmonary - Incapacitating - Nerve - Riot control/tear gas - Vomiting	Liquid/aerosol contaminants can be dispersed using sprayers or other aerosol generators; liquids vaporizing from puddles/containers; or munitions.	Chemicals agents may pose viable threats for hours to weeks, depending on the agent and the conditions in which it exists.	Contamination can be carried out of the initial target area by persons, vehicles, water, and wind. Chemicals may be corrosive or otherwise damaging over time if not remediated.	Air temperature can affect evaporation of aerosols. Ground temperature affects evaporation of liquids. Humidity can enlarge aerosol particles, reducing inhalation hazard. Precipitation can dilute and disperse agents, but can spread contamination. Wind can disperse vapors, but also cause target are to be dynamic. The micro-meteorological effects of buildings and terrain can alter travel and duration of agents. Shielding in the form of sheltering in place can protect people and property from harmful effects.
Conventional Bomb - Stationary vehicle - Moving vehicle - Mail - Supply - Thrown - Placed - Personnel	Detonation of explosive device on or near target; via person, vehicle, or projectile.	Instantaneous; additional secondary devices may be used, lengthening the time duration of the hazard until the attack site is determined to be clear.	Extent of damage is determined by type and quantity of explosive. Effects generally static other than cascading consequences, incremental structural failure, etc.	Energy decreases logarithmically as a function of distance from seat of blast. Terrain, forestation, structures, etc., can provide shielding by absorbing and/or deflecting energy and debris. Exacerbating conditions include ease of access to target; lack of barriers/shielding; poor construction; and ease of concealment of device.
Cyberterrorism	Electronic attack using one computer system against another.	Minutes to days.	Generally no direct effects on built environment.	Inadequate security can facilitate access to critical computer systems, allowing them to be used to conduct attacks.

Source: FEMA

Table 6-1. Event Profiles for Terrorism and Technological Hazards (*Continued*)

Hazard/Threat	Application Mode	Hazard Duration	Extent of Effects; Static/Dynamic	Mitigating and Exacerbating Conditions
Hazardous Material Release (fixed facility or transportation) - Toxic Industrial Chemicals and Materials (Organic vapors: cyclohexane; Acid gases: cyanogens, chlorine, hydrogen sulfide; Base gases: ammonia; Special cases: phosgene, formaldehyde)	Solid, liquid, and/or gaseous contaminants may be released from fixed or mobile containers.	Hours to days.	Chemicals may be corrosive or otherwise damaging over time. Explosion and/or fire may be subsequent. Contamination may be carried out of the incident area by persons, vehicles, water, and wind.	As with chemical weapons, weather conditions will directly affect how the hazard develops. The micro-meteorological effects of buildings and terrain can alter travel and duration of agents. Shielding in the form of sheltering in place can protect people and property from harmful effects. Non-compliance with fire and building codes as well as failure to maintain existing fire protection and containment features can substantially increase the damage from a hazardous materials release.
Nuclear Device	Detonation of nuclear device underground, at the surface, in the air or at high altitude.	Light/heat flash and blast/shock wave last for seconds; nuclear radiation and fallout hazards can persist for years. Electromagnetic pulse from a high-altitude detonation lasts for seconds and affects only unprotected electronic systems.	Initial light, heat and blast effects of a subsurface, ground or air burst are static and are determined by the device's characteristics and employment; fallout of radioactive contaminants may be dynamic, depending on meteorological conditions.	Harmful effects of radiation can be reduced by minimizing the time of exposure. Light, heat, and blast energy decrease logarithmically as a function of distance from seat of blast. Terrain, forestation, structures, etc., can provide shielding by absorbing and/or deflecting radiation and radioactive contaminants.
Radiological Agents - Alpha - Beta - Gamma	Radioactive contaminants can be dispersed using sprayers/aerosol generators, or by point or line sources such as munitions, covert deposits, and moving sprayers.	Contaminants may remain hazardous for seconds to years, depending on material used.	Initial effects will be localized to site of attack; depending on meteorological conditions, subsequent behavior of radioactive contaminants may be dynamic.	Duration of exposure, distance from source of radiation, and the amount of shielding between source and target determine exposure to radiation.

Source: FEMA

Table 6-1. Event Profiles for Terrorism and Technological Hazards (*Continued*)

Hazard/Threat	Application Mode	Hazard Duration	Extent of Effects; Static/Dynamic	Mitigating and Exacerbating Conditions
Surveillance - Acoustic - Electronic eavesdropping - Visual	Stand-off collection of visual information using cameras or high powered optics, acoustic information using directional microphones and lasers, and electronic information from computers, cell phones, and hand-held radios. Placed collection by putting a device "bug" at the point of use.	Usually months.	This is usually the prelude to the loss of an asset. A terrorist surveillance team spends much time looking for vulnerabilities and tactics that will be successful. This is the time period that provides the best assessment of threat as it indicates targeting of the facility.	Building design, especially blocking lines of sight and ensuring the exterior walls and windows do not allow sound transmission or acoustic collection, can mitigate this hazard.
Unauthorized Entry - Forced - Covert	Use of hand or power tools, weapons, or explosives to create a man-sized opening or operate an assembly (such as a locked door), or use false credentials to enter a building.	Minutes to hours, depending upon the intent.	If goal is to steal or destroy physical assets or compromise information, the initial effects are quick, but damage may be long lasting. If intent is to disrupt operations or take hostages, the effects may last for a long time, especially if injury or death occurs.	Standard physical security building design should be the minimum mitigation measures. For more critical assets, additional measures, like closed circuit television or traffic flow that channels visitors past access control, aids in detection of this hazard.

Source: FEMA

These ratings may be changed over time. What may be a high threat in the present may, over time, be lessened to a medium or low threat depending upon the conditions at a particular time.

Alternative Approach

Assessing terrorist threats is the most difficult aspect of planning to resist terrorist attacks. This is particularly true for those building owners and facility managers who have not had any experience in doing so. An effective alternative approach may be to select a level of desired protection for a business operation based on management decision making, and then proceeding to a vulnerability assessment. Various federal agencies along with the U.S. Department of Defense correlate "levels of protection" with potential damage and expected injuries. The following levels are based on Department of Defense definitions:

- **High Protection**. Facility superficially damaged; no permanent deformation of primary and secondary structural members or non-structural elements. Only superficial injuries are likely.

- **Medium Protection**. Damaged, but repairable. Minor deformations of non-structural elements and secondary structural members and no permanent deformation in primary structural members. Some minor injuries, but fatalities are unlikely.

- **Very Low Protection**. Heavily damaged, onset of structural collapse. Major deformation of primary and secondary structural members, but progressive collapse is unlikely. Collapse of non-structural elements. Majority of personnel suffer serious injuries. There are likely to be a limited number—10 percent to 25 percent—of fatalities.

Note that the "very low" level is not the same as doing nothing. No action could result in catastrophic building failure and high loss of life.

VULNERABILITY ASSESSMENT

A terrorism vulnerability assessment evaluates any weaknesses that can be exploited by a terrorist. It evaluates the vulnerability of facilities across a broad range of identified threats/hazards and provides a basis for determining physical and operational mitigation measures for their protection. It applies both to new building programming and design as well as to existing building management and renovation over the service life of a structure.

A vulnerability rating can also be assigned to the appropriate aspects of building operations and systems to the defined threats for the particular facility. These ratings can also be assigned as high, medium or low:

- **High Vulnerability**. One or more significant weaknesses have been identified that make the facility highly susceptible to a terrorist or hazard.

- **Medium Vulnerability**. A weakness has been identified that makes the facility somewhat susceptible to a terrorist or hazard.

- **Low Vulnerability**. A minor weakness has been identified that slightly increases the susceptibility of the facility to a terrorist or hazard.

Initial Vulnerability Estimate

The initial vulnerability estimate provides a quick, qualitative assessment of the vulnerability of existing buildings to terrorist attack. Three means of data collection using a simple scale of high, medium and low ratings may provide useful information. The data collection is based upon three criteria:

- Visual inspection
- Document review
- Organization and management procedures review

Visual Inspection

When visually inspecting the condition of the property, an evaluation of the site and all the facility systems is performed. This includes the architectural, structural, building envelope, utility, mechanical, plumbing and gas, electrical, fire alarm, communications and information technology systems. Equipment operations and maintenance procedures and records, security systems, and planning and procedures should also be evaluated. The inspection may need to go beyond the site to determine the vulnerability of utility and other infrastructure systems.

Document Review

The planning team, which includes the building owner and the facility manager, should review all necessary plans, specifications and related construction data in terms of terrorism vulnerability. Equipment operation, maintenance procedures and records, as well as security procedures, should be included in this review.

Organization and Management Procedures Review

The planning team should review business and operations practices and procedures to identify opportunities that can reduce exposure to attack. This review also includes tenant operations.

Vulnerability Estimate Screening

The following screening table from FEMA provides guidance for initial vulnerability assessment. The goal of this assessment is to distinguish facilities of high, medium or low vulnerability to terrorist attack. The implication is that high vulnerability facilities should receive more detailed analysis. Specific strategies for risk reduction should be developed.

For this initial assessment, subjective ratings by building owners, facility managers and other qualified professionals who are familiar with the facility, are appropriate. Assigning a "high," "medium" or "low" vulnerability rating to the responses to vulnerability questions for each building system, will provide a preliminary basis for estimating the overall vulnerability of a particular facility to terrorist attack. The responses will also indicate areas of opportunity for mitigation actions to reduce terrorism risk.

Site Questions

Vulnerability assessment of the "site" examines surrounding structures, terrain, perimeter controls, traffic patterns and separations, landscaping elements/features, and lines of sight.

"Site" questions focus primarily on visual inspection to develop ratings. The questions emphasize vulnerability to moving vehicle, stationary vehicle, and covert entry tactics. Vulnerability to blast is the primary concern addressed (see Table 6-2).

Architectural Questions

Assessing "architectural" vulnerability investigates tenancy, services, public and private access, access controls, activity patterns and exposures.

"Architectural" questions focus equally on visual inspection and evaluation of organizational and management procedures to develop ratings. The questions emphasize vulnerability to moving vehicle, stationary vehicle and covert entry tactics. Vulnerability to blast is a primary concern (see Table 6-3).

Structural and Building Envelope Questions

A vulnerability assessment of "structural" systems examines construction type, materials, detailing, collapse characteristics and critical elements. An assessment of the "building envelope" examines strength, fenestration, glazing characteristics/detailing and anchorage.

These questions rely on a review of construction documents and visual inspection to develop ratings. Vulnerability to blasts is the primary concern (see Table 6-4).

Table 6-2. Site Systems Vulnerability Estimate

	Vulnerability Rating (H, M, L)	Visual inspection	Document review	Org/Mgmt procedure	Moving vehicle	Stationary vehicle	Covert entry	Mail	Supplies	Blast effects	Airborne (contamination)	Waterborne (contamination)
What major structures surround the facility?	❑	●	●									
What critical infrastructure, government, military, or recreation facilities are in the local area that impact transportation, utilities, and collateral damage (attack at this facility impacting the other major structures or attack on the major structures impacting this facility)?	❑	●	●	●								
What are the adjacent land uses immediately outside the perimeter of this facility?	❑	●	●									
What are the site access points to the facility?	❑	●			●		●					
What is the minimum distance from the inspection location to the building?	❑	●			●		●			●		
Is there any potential access to the site or facility through utility paths or water runoff?	❑	●	●				●					
What are the existing types of vehicle anti-ram devices for the facility?	❑	●			●					●		
What is the anti-ram buffer zone standoff distance from the building to unscreened vehicles or parking?	❑	●			●							
Are perimeter barriers capable of stopping vehicles?	❑	●	●		●							
Does site circulation prevent high-speed approaches by vehicles?	❑	●			●							
Is there a minimum setback distance between the building and parked vehicles?	❑	●				●				●		
Does adjacent surface parking maintain a minimum standoff distance?	❑	●				●				●		
Do site landscaping and street furniture provide hiding places?	❑	●					●					

LEGEND: ❑ = Determine high, medium, or low vulnerability rating. ● = Applicability of factor to question.

Source: FEMA

Utility Systems Questions

A vulnerability assessment of "utility" systems examines the full range of source and supply systems serving the facility including water, fuel and the electricity supply, as well as the fire alarm and suppression systems and communication systems (see Table 6-5).

These questions rely on information obtained from visual inspection, review of construction documents and organizational/management procedures to develop the ratings. Vulnerability to waterborne contaminants is the primary consideration (see Table 6-5).

Table 6-3. Architectural Systems Vulnerability Estimate

	Vulnerability Rating (H, M, L)	Visual inspection	Document review	Org/Mgmt procedure	Moving vehicle	Stationary vehicle	Covert entry	Mail	Supplies	Blast effects	Airborne (contamination)	Waterborne (contamination)
What major structures surround the facility?	❑	●	●	●								
Do entrances avoid significant queuing?	❑	●		●								
What are the adjacent land uses immediately outside the perimeter of this facility?	❑	●		●								
Are public and private activities separated?	❑	●					●					
Are critical assets (people, activities, building systems and components) located close to any main entrance, vehicle circulation, parking, maintenance area, loading dock, or interior parking?	❑	●	●	●	●	●	●			●		
Are high-value or critical assets located as far into the interior of the building as possible and separated from the public areas of the building?	❑	●	●	●						●		
Is high visitor activity away from critical assets?	❑			●			●					
Are critical assets located in spaces that are occupied 24 hours per day?	❑			●								
Are assets located in areas where they are visible to more than one person?	❑			●								
Do interior barriers differentiate level of security within a facility?	❑	●	●	●								
Are emergency systems located away from high-risk areas?	❑	●	●	●								

LEGEND: ❑ = Determine high, medium, or low vulnerability rating. ● = Applicability of factor to question.

Source: FEMA

Mechanical Systems

A vulnerability assessment of mechanical systems examines air supply and exhaust configurations, filtration, sensing and monitoring, system zoning and control, and elevator management.

These questions and ratings rely on information obtained from review of construction documents and visual inspection. Vulnerability to airborne contaminants is the primary concern, including contamination from chemical, biological and radiological attack (see Table 6-6).

Plumbing and Gas Systems

A vulnerability assessment of plumbing and gas systems examines the liquid distribution systems serving the facility including water and fuel distribution, water heating and fuel storage.

These questions rely on information from a review of construction

Table 6-4. Structural and Building Envelope Systems Vulnerability Estimate

	Vulnerability Rating (H, M, L)	Visual inspection	Document review	Org/Mgmt procedure	Moving vehicle	Stationary vehicle	Covert entry	Mail	Supplies	Blast effects	Airborne (contamination)	Waterborne (contamination)
What type of construction?	❑	•	•							•		
Is the column spacing minimized so that reasonably sized members will resist the design loads and increase the redundancy of the system?	❑	•	•							•		
What are the floor-to-floor heights?	❑	•	•							•		
Is the structure vulnerable to progressive collapse?	❑	•	•									
Are there adequate redundant load paths in the structure?	❑	•	•							•		
What is the designed or estimated protection level of the exterior walls against the postulated explosive threat?	❑		•							•		

LEGEND: ❑ = Determine high, medium, or low vulnerability rating. • = Applicability of factor to question.

Source: FEMA

documents to develop ratings. Vulnerability to waterborne contaminants is the primary concern (see Table 6-7).

Electrical Systems

A vulnerability assessment of electrical systems examines transformer and switchgear security, electricity distribution and accessibility, as well as emergency systems.

These questions rely on information from visual inspection and a review of construction documents to develop ratings. No particular attack mechanism is emphasized (see Table 6-8).

Fire Alarm Systems

A vulnerability assessment of fire alarm systems examines detection sensing and signaling, system configurations, accessibility of controls and redundancies.

These questions rely on information obtained from the review of the construction documents, as well as a review of organizational/management procedures to develop ratings. No particular attack mechanism is emphasized (see Table 6-9).

Table 6-5. Utility Systems Vulnerability Estimate

	Vulnerability Rating (H, M, L)	Visual inspection	Document review	Org/Mgmt procedure	Moving vehicle	Stationary vehicle	Covert entry	Mail	Supplies	Blast effects	Airborne (contamination)	Waterborne (contamination)
What is the source of domestic water? (utility, municipal, wells, lake, river, storage tank)	❑	●	●									●
How many gallons and how long will it allow operations to continue?	❑	●	●	●								●
What is the source of water for the fire suppression system? (local utility company lines, storage tanks with utility company backup, lake, or river)	❑	●	●									
Are there alternate water supplies for fire suppression?	❑	●	●	●								
Are the sprinkler and standpipe connections adequate and redundant?	❑	●	●									
What fuel supplies do the facility rely upon for critical operation?	❑	●	●	●								
Where is the fuel supply obtained?	❑			●								
Are there alternate sources of fuel?	❑			●								
Can alternate fuels be used?	❑		●	●								
What is the normal source of electrical service for the facility?	❑	●	●									
What provisions for emergency power exist? What systems receive emergency power and have capacity requirements been tested?	❑	●	●	●								
By what means does the main telephone and data communications interface the facility?	❑	●	●	●								

LEGEND: ❑ = Determine high, medium, or low vulnerability rating. ● = Applicability of factor to question.

Source: FEMA

Communications and Information Technology Systems

A vulnerability assessment of communications and information technology systems examines distribution, power supplies, accessibility, control, notification and backups.

These questions rely on information from visual inspection, a review of construction documents, as well as a review of the organizational/management procedures to develop ratings. No particular attack mechanism is emphasized (see Table 6-10).

As companies and organizations increase their security measures and attempt to identify vulnerabilities in critical assets, they are looking for a mechanism to ensure an efficient investment of resources to counter threats. One such mechanism is a risk management model that will

Table 6-6. Mechanical Systems Vulnerability Estimate

	Vulnerability Rating (H, M, L)	Visual inspection	Document review	Org/Mgmt procedure	Moving vehicle	Stationary vehicle	Covert entry	Mail	Supplies	Blast effects	Airborne (contamination)	Waterborne (contamination)
Where are the air intakes and exhaust louvers for the building? (low, high, or midpoint of the building structure)	❑	●	●									●
Are there multiple air intake locations?	❑	●	●									●
How are air handling systems zoned?	❑	●	●									●
Are there large central air handling units or are there multiple units serving separate zones?	❑		●									●
Are there any redundancies in the air handling system?	❑		●	●								●
Where is roof-mounted equipment located on the roof? (near perimeter, at center of roof)	❑	●										

LEGEND: ❑ = Determine high, medium, or low vulnerability rating. ● = Applicability of factor to question.

Source: FEMA

Table 6-7. Plumbing and Gas Systems Vulnerability Estimate

	Vulnerability Rating (H, M, L)	Visual inspection	Document review	Org/Mgmt procedure	Moving vehicle	Stationary vehicle	Covert entry	Mail	Supplies	Blast effects	Airborne (contamination)	Waterborne (contamination)
What is the method of water distribution?	❑		●									●
What is the method of gas distribution? (heating, cooking, medical, process)	❑		●									
What is the method of heating domestic water?	❑	●	●	●								
Are there reserve supplies of critical gases?	❑		●	●								

LEGEND: ❑ = Determine high, medium, or low vulnerability rating. ● = Applicability of factor to question.

Source: FEMA

- assess assets, threats and vulnerabilities; and
- incorporate a continuous assessment feature.

The questions listed above provide a framework for such a model. By reviewing these ratings, a preliminary determination can be made of where the major vulnerabilities and threats could occur in a facility.

Table 6-8. Electrical Systems Vulnerability Estimate

	Vulnerability Rating (H, M, L)	Visual inspection	Document review	Org/Mgmt procedure	Moving vehicle	Stationary vehicle	Covert entry	Mail	Supplies	Blast effects	Airborne (contamination)	Waterborne (contamination)
Are there any transformers or switchgears located outside the building or accessible from the building exterior?	❑	●										
Are they (transformers or switchgears) vulnerable to public access?	❑	●										
Are critical electrical systems located in areas outside of secured electrical areas?	❑	●	●	●								
Does emergency backup power exist for all areas within the facility or for critical areas only?	❑	●	●									

LEGEND: ❑ = Determine high, medium, or low vulnerability rating. ● = Applicability of factor to question.

Source: FEMA

Table 6-9. Fire Alarm Systems Vulnerability Estimate

	Vulnerability Rating (H, M, L)	Visual inspection	Document review	Org/Mgmt procedure	Moving vehicle	Stationary vehicle	Covert entry	Mail	Supplies	Blast effects	Airborne (contamination)	Waterborne (contamination)
Is the fire alarm system stand-alone or integrated with other functions such as security and environmental or building management systems?	❑		●	●								
Is there redundant off-premises fire alarm reporting?	❑		●	●								

LEGEND: ❑ = Determine high, medium, or low vulnerability rating. ● = Applicability of factor to question.

Source: FEMA

Additionally, a preliminary determination can be made regarding which risks require immediate attention. As noted, however, the ratings (low, medium, high) are subjective and as a result may not be exact. A more detailed and quantitative evaluation is required. This detailed evaluation involves a significantly more thorough review of information in all areas, including additional information that concerns:

- Equipment Operations and Maintenance (up-to-date drawings, manuals and procedures, training, monitoring, etc.);

Table 6-10. Communication and its Systems Vulnerability Estimate

	Vulnerability Rating (H, M, L)	Visual inspection	Document review	Org/Mgmt procedure	Moving vehicle	Stationary vehicle	Covert entry	Mail	Supplies	Blast effects	Airborne (contamination)	Waterborne (contamination)
Where is the main telephone distribution room and where is it in relation to higher risk areas?	❑	•	•	•								
Where are communication systems wiring closets located? (voice, data, signal, alarm)	❑	•	•									

LEGEND: ❑ = Determine high, medium, or low vulnerability rating. • = Applicability of factor to question.

Source: FEMA

- Security Systems (perimeter and interior sensing, monitoring and control, security system documentation and training, etc.); and

- The Security Master Plan (currency, responsibilities, etc.).

These questions should be considered when fully evaluating vulnerability to terrorist threats. The means of data collection that should be employed and the particular tactics and attack mechanisms addressed by each question are identified in the appendix. This is done so that specialized checklists can be created to assess vulnerability to terrorist tactics of particular concern to an individual company or organization.

By following such a model, organizations are able to "tailor" their management of risk to the current situation, as well as to assess future risks.

Sources

Federal Emergency Management Agency (FEMA), *Risk Management Series: Insurance, Finance, and Regulation Primer for Terrorism Risk Management in Buildings*, FEMA 429. December 2003.

Federal Emergency Management Agency (FEMA), *Risk Management Series: Primer for Design of Commercial Buildings to Mitigate Terrorist Attacks*, FEMA 427. December 2003.

Gustin, Joseph, *Cyber Terrorism: A Guide for Facility Managers*, Lilburn, GA: The Fairmont Press, Inc., 2004.

Gustin, Joseph F., *Disaster and Recovery Planning: A Guide for Facility Managers, 4th ed.*, Lilburn, GA: The Fairmont Press, Inc., 2007.

National Infrastructure Protection Center (NIPC). *Risk Management: An Essential Guide to Protecting Critical Assets*, November 2002.

Chapter 7

Violence In The Workplace

Homicides, robberies, beatings, sexual assaults—sounds like crime stories on life in an urban America broadcast on the News at 11, right? These are also the types of violent acts that occur in the workplace far too often. Although a well-planned and carefully developed violence prevention program cannot address every potentially violent incident, facility managers can minimize the risk of violence and its serious consequences.

CRIME IN THE WORKPLACE: AN OVERVIEW

The Bureau of Labor Statistics (BLS) reported that the number of workplace homicides in 2005 was higher than in 2004—there were 564 workplace homicides recorded in 2005 showing a slight increase from the 559 recorded in 2004. According to the BLS, workplace suicides were sharply lower in 2005. There were a total of 177 suicides in 2005 reflecting a 14 percent decline.

Table 7-1 describes the four most frequent work-related fatal events from 1992-2005. Since 1994, workplace homicides have declined. In 1994, workplace homicides were the leading cause of death. In 2005, however, workplace homicides fell to fourth place in 2005 (see Figure 7-1.)

According to the BLS, assaults and violent acts accounted for fourteen percent of the total number of workplace fatalities in 2005. And ten percent of these fatalities included homicides (see Table 7-2).

Women had a higher percentage of fatal injuries resulting from highway incidents and homicides than men, while men had a higher percentage of fatal work injuries from falls and from contact with objects and equipment (see Table 7-2).

In December, 2001, The Bureau of Justice Statistics, National Crime Victimization Survey on Violence in the Workplace from 1993-2000, released a report that found that approximately 1.7 million violent workplace crimes were committed. Additionally, during that period an

Table 7-1. The four most frequent work-related fatal events, 1992-2005.

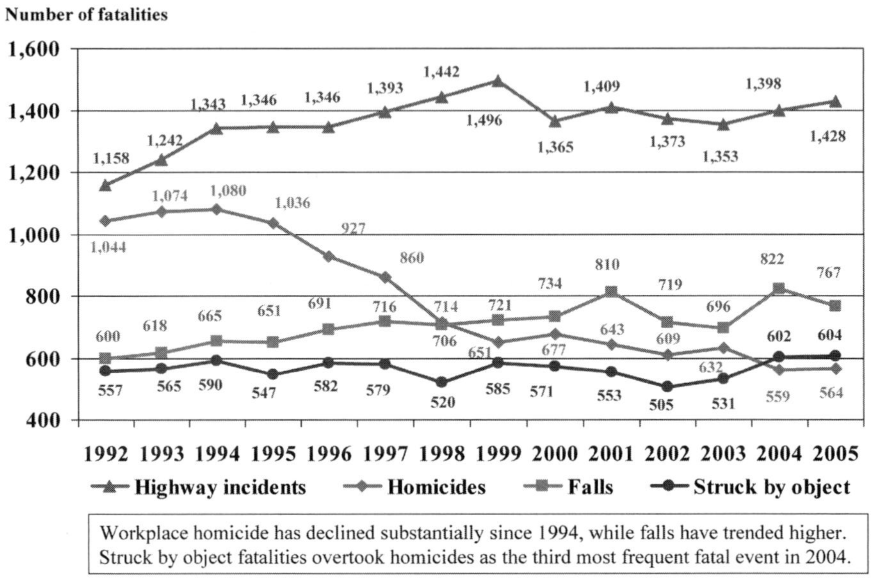

NOTE: Data from 2001 exclude fatalities resulting from the September 11 terrorist attacks.
SOURCE: US Department of Labor, Bureau of Labor Statistics, Census of Fatal Occupational Injuries, 2005.

average of 900 homicides also occurred. With those statistics in mind, it should be noted that these numbers represent a *decrease* in workplace violent crime. Since 1993, workplace homicides have been on the declines. The FBI reports that in the early 1990s a peak of 1000 workplace homicides were reported compared to approximately 677 in the year 2000. The majority of these workplace homicides, about 77%, resulted from robberies and related crimes.

The remaining homicides were committed by disgruntled employees, clients/customers. Also accounting for workplace homicides were the domestic violence/stalking actions that carry over into the workplace.

Company/industry awareness of the violence in their affected industries is one of the significant reasons that there is a decline in workplace homicides. Implementation of safety and security measures such as bullet proof glass, improved lighting, video cameras, etc., by the company management can account for the decline in workplace homicides from the late 1980s-1990s.

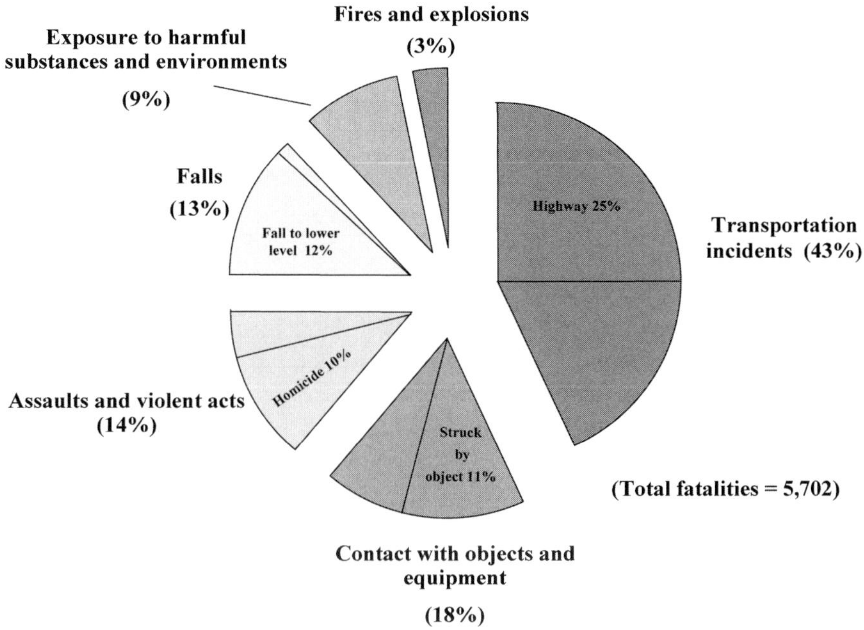

More work-related fatalities resulted from transportation incidents than from any other event. Highway incidents alone accounted for nearly one out of every four fatal work injuries in 2005.

NOTE: Percentages may not add to totals of rounding.
SOURCE: US Department of Labor, Bureau of Labor Statistics, Census of Fatal Occupational Injuries, 2005.

Figure 7-1. The manner in which workplace fatalities occurred, 2005.

Categories of Workplace Violence

The FBI classifies workplace violence into four broad categories which include:

1. Violence committed by criminals who enter to commit robbery or another crime and who have no connection with the workplace;

2. Violence directed at employees by customers, clients, patients, etc., for whom an organization provides service;

Table 7-3. Fatal work injury incidents varied between men and women, 2005

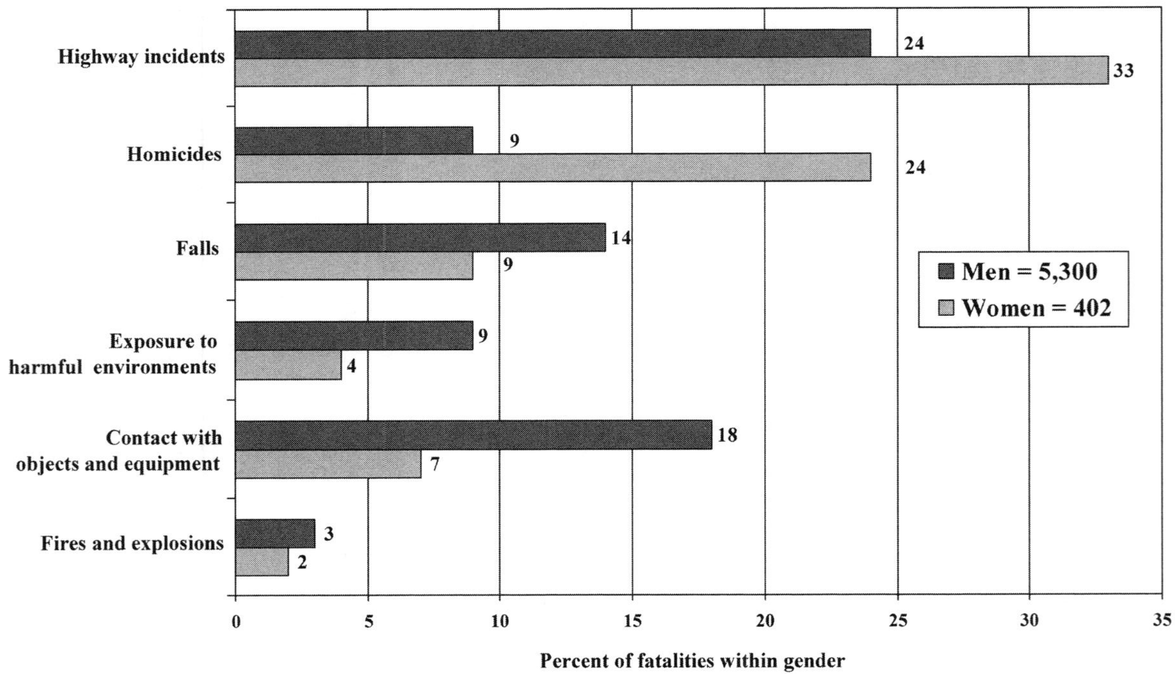

Women had a higher percentage of fatal injuries resulting from highway incidents and homicides than men, while men had a higher percentage of fatal work injuries from falls and from contact with objects and equipment.

NOTE: Percentages do not add to 100% because not all categories are shown.
SOURCE: US Department of Labor, Bureau of Labor Statistics, Census of Fatal Occupational Injuries, 2005.

3. Violence against co-workers, supervisors or managers by a present or former employee; and,

4. Violence committed in the workplace by someone who is personally involved with an employee such as an abusive spouse, domestic partner, boyfriend or girlfriend, etc.

Robbery is the prime motivator for the majority of workplace homicides. The common idea that workplace homicides are committed by co-workers, clients or spouses—employees going "postal"—is false. Only one in seven homicide victims were killed by a co-worker or personal associate.

Nearly one half of all homicide victims were employed in retail establishments (mainly eating and drinking places) and grocery stores. Younger workers were more likely to be victims of homicide than older workers, with persons between the ages of 16 and 17 leading the mortality figures (see Table 7-3).

In 2004, there were a total of 559 homicides of which 421 or 75% were the result of shootings.

These alarming statistics have sobering consequences for everyone. Most significant, of course, is the needless loss of human life. Companies and their employees are thrown into a tailspin when the senseless acts of violence occur. From an economic perspective, companies lose millions of dollars in lost productivity and downtime. A study of the effects of workplace violence conducted by the U.S. Department of Justice reveal that workplace violence costs employers millions of lost work days each year. These lost work days translate into millions of dollars in lost wages. Also impacted by workplace violence are the legal and security costs, medical costs and worker's compensation costs.

From a human perspective, the toll is even more exacting. Employees who are the victims of violence, as well as their co-workers, can and often do ride an emotional roller coaster. Reaction to violence in the workplace can manifest itself in various ways.

Workers can and often do experience heightened anxiety as well as depression. Anger—itself a significant factor in violence, nightmares, sleeplessness, withdrawal, paranoia and, in some cases, substance abuse, is a common response to violence-induced stress.

Table 7-3. Fatal occupational injuries by selected worker characteristics and selected event or exposure, 2005.

Characteristic	Fatalities		Selected event or exposure[1] (percent of total for characteristic category)			
	Number	Percent	Highway[2]	Homicides	Falls	Struck by object
Total	5,702	100	25	10	13	11
Employee status						
Wage and salary[3]	4,568	80	28	9	14	10
Self-employed[4]	1,134	20	13	14	12	15
Sex						
Men	5,300	93	24	9	14	11
Women	402	7	33	24	9	4
Age[5]						
Under 16 years	24	([6])	–	–	–	12
16-17 years	30	1	13	13	10	20
18-19 years	112	2	21	10	16	11
20-24 years	403	7	26	10	10	9
25-34 years	1,005	18	24	12	11	10
35-44 years	1,239	22	26	12	12	11
45-54 years	1,383	24	26	9	14	10
55-64 years	924	16	28	8	16	10
65 years and older	575	10	20	6	17	15
Race or ethnic origin[7]						
White	3,940	69	26	7	13	11
Black or African-American	577	10	29	20	8	8
Hispanic or Latino	917	16	19	10	20	12
American Indian or Alaska Native	49	1	31	–	8	10
Asian	153	3	12	46	9	6
Native Hawaiian or Pacific Islander	9	([6])	–	–	–	–
Other or not reported	55	1	15	25	11	7

[1] The figure shown is the percent of the total fatalities for that demographic group.
[2] "Highway" includes deaths to vehicle occupants resulting from traffic incidents that occur on the public roadway, shoulder, or surrounding area. It excludes incidents occurring entirely off the roadway, such as in parking lots and on farms; incidents involving trains; and deaths to pedestrians or other nonpassengers.
[3] May include volunteers and workers receiving other types of compensation.
[4] Includes self-employed workers, owners of unincorporated businesses and farms, paid and unpaid family workers, members of partnerships, and may include owners of incorporated businesses.
[5] There were 7 fatalities for which there was insufficient information to determine the age of the decedent.
[6] Less than or equal to 0.5 percent.
[7] Persons identified as Hispanic or Latino may be of any race. The race categories shown exclude Hispanic and Latino workers.
NOTE: Totals for 2005 are preliminary. Totals for major categories may include subcategories not shown separately. Percentages may not add to totals because of rounding. Dashes indicate no data reported or data that do not meet publication criteria.
SOURCE: U.S. Department of Labor, Bureau of Labor Statistics, in cooperation with State, New York City, District of Columbia, and Federal agencies, Census of Fatal Occupational Injuries

EMPLOYER RESPONSE

Since the issue of workplace violence is first and foremost a safety issue, facility managers must be prepared to assist employers in responding to the issue. However, before a response can be prepared, several questions need to be answered.

Since there is an Occupational Safety and Health Administration (OSHA) requirement that employers are required to provide to provide "a safe and healthful working environment" for their employees the question is—what are employers required to do in order to provide this "safe and healthful working environment?"

What can employers do to prevent violence from occurring?

What should employers do to assist employees—victims as well as co-workers—to cope with the after-effects of the violent act?

Employer Requirements

There are no specific regulatory standards that govern workplace violence. Two significant "actions," however, have taken place. The Secretary of Labor has directed OSHA to study the issue of workplace violence to determine why more than 8,000 workers have been killed since 1980. And OSHA has issued safety and security guidelines for health care and community services workers.

These guidelines can suggest the possible direction that employers will be required to take regarding workplace violence. *The Guidelines for Security and Safety and Health Care and Community Service Workers* stresses the need for a written program for job, safety, health and security.

The Written Safety Program

A written safety program should be developed and incorporated into the Disaster and Recovery Plan. This written safety program, depending upon the size of the company, begins with a planning group. This planning group should evaluate the company's current capability to response to incidents of workplace violence. Once evaluated, recommendations for response improvement should be defined.

Who should be included in this planning group? Again, depending upon the size of the company, representatives from management, Employee Relations, Employee Assistance, Facility, Security, Public Relations, Employee Unions and Personnel should be involved. In smaller companies, the Facility manager alone may be responsible. In any event, the company's legal department should also be consulted once recommendations are finalized.

Once the planning group is formed, what are the steps involved in the planning process? The General Services Administration suggests the following:

1. Analyze the company's current ability to handle potentially violent situations.

2. Fill the skills gap.

3. Develop an employee-reporting procedure.

4. Develop a response plan

The written safety program, which begins with analyzing the worksite, includes reviewing a company's client type, physical facility/site, staffing levels, lighting, accessibility to the physical facility and prior security problems. A written safety program has several advantages including (according to the GSA):

- Informing employees what the policy covers;
- Encouraging employees to report incidents;
- Informing employees whom to call; and
- Demonstrating senior management's commitment in dealing with reported incidents.

The program's effectiveness is further enhanced by securing top management and employee commitment and involvement. Other core components of the written safety program include involving law enforcement agencies and the ongoing training and education of all employees.

The Policy Statement

The workplace violence policy statement should include the following:

- The types of offenses that contribute to workplace violence: physical violence, harassment, intimidation and other disruptive behavior;
- The types of incidents involving individuals: employee to employee, and client/customer/non-employee to employee incidents;
- The responsibility of employees for maintaining a safe work environment;
- The company response to all reported incidents;
- The company actions to stop inappropriate behavior; and
- The assurance of management support.

The success of the violence prevention program rests with all employees—management as well as "rank and file"—to recognize the potential for violence and knowing how to respond to the violent act.

Recognizing the Warning Signs of Violence

There are certain behavioral characteristics that people with a propensity towards violence share. Given the "right" setting, individuals who fit the "profile" of a violence-prone individual may react to a situation, or set of circumstances, violently. While analyzing past incidents of workplace violence, The Federal Bureau of Investigation's (FBI)'s National Center for the Analysis of Violent Crime (NVAVC), Profiling and Behavior Assessment Unit identified some of the indicators of increased risk of violent behavior. These include:

- Direct or veiled threats of harm;
- Intimidating, belligerent, harassing, bullying, or other inappropriate and aggressive behavior.
- Numerous conflicts with supervisors and other employees;
- Bringing in a weapon to the workplace, brandishing a weapon in the workplace; making inappropriate references to guns, or fascination with weapons;
- Statements made by individuals:
  - Showing fascination with incidents or workplace violence,
  - Indicating approval of the use of violence to resolve a problem,
  - Indicating identification with perpetrators of workplace homicides;
  - Indicating desperation (over family, financial, and other personal problems) to the point of contemplating suicide;
- Drug/alcohol abuse; and
- Extreme changes in behavior.

An employee facing such problems as a job layoff or termination, stressful family situations such as a death, divorce, etc., or employees suffering from mental illness or who are in therapy, are also indicators that can lead someone into physical violence.

However, it is important to keep in mind that employers should not automatically presume that a person "fitting" the profile will commit a violent act. About 20 percent of individuals who fit the profile are not any more likely to be violent than most people.

WORKPLACE VIOLENCE PREVENTION—RESPONSIBILITIES

Who is responsible for the safety of employees and staff of a company/industry? All employees involved share in the responsibility for maintaining a safe work environment. The following section, outlined by the United States Department of Agriculture (USDA), provides a more detailed description of the responsibilities of various persons or offices including:

- Employees
- Managers and Supervisors
- Department Heads
- Human Resources Staff
- Employee Assistance Program Counselors
- Union Employee Organizations
- Facilities Staff
- Security Staff
- Law Enforcement Staff
- Conflict Resolution Offices

Employee Responsibilities

- Be familiar with the company policy regarding workplace violence.
- Be responsible for securing their own workplace.
- Be responsible for questioning and/or reporting strangers to supervisors.
- Be aware of any threats, physical or verbal, and/or any disruptive behavior of any individual and report such to supervisors.

- Be familiar with the procedures for dealing with workplace threats and emergencies.
- Do not confront threatening individuals.
- Take all threats seriously.

Managers and Supervisor Responsibilities

- Inform employees of company policies and procedures.
- Ensure that employees know specific procedures for dealing with workplace threats and emergencies including how to contact police, fire and other safety and security personnel.
- Ensure that employees with special needs are aware of emergency evacuation procedures and have assistance (as necessary) regarding emergency evacuation situations.
- Respond to potential threats and escalating situations by utilizing proper resources from the following: local law enforcement and medical services, human resources staff and the Employee Assistance Program, if available.
- Take all threats seriously.
- Check prospective employees' background prior to hiring.

Department Head Responsibilities

- Develop a policy statement reflecting that the company will not tolerate violent or disruptive behavior and that all reported incidents will be taken seriously.
- Ensure that the company's workplace prevention policy is available to employees.
- Ensure that employees are aware of the policies, procedures and instructions in the company policy.
- Ensure that safety and law enforcement personnel have completed an on-site review of the safety and security of the building and offices.

- Provide adequate resources for employee training and awareness.

- Ensure that performance standards of appropriate staff reflect the importance of workplace safety and security.

- Provide for briefings on workplace violence at staff meetings.

Human Resources Staff Responsibilities

- Provide for supervisory training which includes basic leadership skills such as:
  - — Setting clear standards of conduct and performance.
  - — Addressing employee problems promptly.
  - — Using the probationary periods, performance counseling, discipline, and other management tools conscientiously.

- Provide technical expertise and consultation to help supervisors determine what course of administrative action is most appropriate in specific situations.

- Determine whether disciplinary action should be taken.

- Help supervisors determine proper reasonable accommodation.

Employee Assistance Program Counselor Responsibilities

- Provide short-term counseling and referral services to employees at no cost.

- Help in the prevention of workplace violence through:
  - — Early involvement in organizational change.
  - — Training employees in dealing with angry co-workers and customers; conflict resolution and communication skills
  - — Training supervisors to deal with problems as soon as they surface without diagnosing the employee's problem.
  - — Consultation with supervisors to identify specific problem areas, develop action plans to resolve problems in the early stages, and encourage employees to contact the EAP, if available, for individual counseling.
  - — Consultation with incident response teams when a potential for violence exists or an actual incident is reported.

— Participation on critical incident stress debriefing teams in the event of a violent situation.

Union/Employee Organization Responsibilities

- Be familiar with and actively support policy and contract language on workplace violence prevention.
- Stay alert to security issues and potential threats.
- Be aware of procedures for addressing workplace threats and emergencies.
- Be aware of the Employee Assistance Program including the procedures/policy regarding the ability of designated union officials to make employee referrals to the EAP.
- Work closely with all levels of management to ensure that employees are up-to-date on company workplace violence prevention policy and procedures.
- Participate fully with management in all phases of workplace violence prevention and response, including membership on threat assessment and incident response teams.

Facilities Staff/Securities Staff Responsibilities

- Serve as the liaison with law enforcement as well as the company expert on security matters.
- Conduct regular threat assessment survey of the facility to determine the level of security preparedness and any gaps in the security posture.
- Serve as the facility security expert, keeping management advised of the risk of violence, the security gaps identified by threat assessments, and the means to close these gaps, including the latest technologies.
- Work with facility personnel to improve the security levels of the buildings, grounds, parking lots, etc.
- Train facility personnel in security measures and violence prevention techniques.

Facilities personnel should work closely with security staff to ensure that the buildings, areas and grounds are safe for employees and visitors.

This includes not only keeping buildings and grounds well maintained but participating with security personnel in threat assessment surveys, keeping management informed of the status of the physical plan, and providing budget request with justification for security upgrades.

Law Enforcement Staff Responsibilities

- Identify in advance the types of situations that may occur and when and how law enforcement should be notified of an incident.
- Indicate whether law enforcement officers have jurisdictional restrictions and identify alternative law enforcement agencies that may be able to provide assistance.
- Indicate whether law enforcement officers have arrest authority.
- Provide threat assessment personnel who can assist the company in determining the best way to protect personnel.
- Suggest safety and security measures that need to be implemented.
- Arrange for all-employee briefings or training on specific workplace violence issues.

Workplace Violence Prevention—Employee Training

An effective workplace violence prevention training program that emphasizes management's commitment to employee safety is an important component of a company's disaster prevention plan. The training program must emphasize that management will take a proactive approach to reported incidents of threats, intimidation, harassment, etc.

The training program should inform employees that:

- Management will take reports of threats seriously;
- Incidents should be reported; and
- Management is committed to deal with the reported incidents.
- The training program should also emphasize that:

- All employees should know how to report incidents of violent, intimidating, threatening and other disruptive behavior.
- All employees should know the procedures for reporting incidents, including the phone number of the appropriate personnel with phone numbers during a crisis or an emergency.
- Additionally, workplace violence prevention training for employees may also include the following topics:
  - — Explanation of the company's workplace violence policy;
  - — Encouragement to report incidents;
  - — Ways of preventing or diffusing volatile situations or aggressive behavior;
  - — How to deal with hostile persons;
  - — Managing anger;
  - — Techniques and skills to resolve conflicts;
  - — Stress management, relaxation techniques, wellness training;
  - — Security procedures, e.g., the location and operation of safety devices such as alarm systems.
- Supervisory training for workplace violence prevention may also include:
  - — Ways to encourage employees to report incidents in which they feel threatened for any reason by anyone inside or outside the organization;
  - — Skills in behaving compassionately and supportively towards employees who report incidents;
  - — Skills in taking disciplinary actions;
  - — Basic skills in handling crisis situations;
  - — Basic emergency procedures, and
  - — How to ensure that appropriate screening of pre-employment references has been done.

Workplace Violence Checklist OSHA

The checklist on the following pages can be used to identify and evaluate workplace security hazards (see Table 7-4). Please note that "TRUE" indicates a potential risk for serious security hazards.

Handling Threats

Employees should be encouraged to report any and all threats made against them to their supervisor. Threats, no matter how in-

Table 7-4. Workplace Violence Checklist. Source: Occupational Safety and Health Administration (OSHA)

T F	This industry frequently confronts violent behavior and assaults of staff.
T F	Violence has occurred on the premises or in conducting business.
T F	Customers, clients, or coworkers assault, threaten, yell, push, or verbally abuse employees or use racial or sexual remarks.
T F	Employees are **NOT** required to report incidents or threats of violence, regardless of injury or severity to the employer.
T F	Employees have **NOT** been trained by the employer to recognize and handle threatening aggressive or violent behavior.
T F	Violence is accepted as "part of the job" by some managers, supervisors, and/or employees.
T F	Access and freedom of movement within the workplace are **NOT** restricted to those persons who have a legitimate reason for being there.
T F	The workplace security system is inadequate—i.e., door locks malfunction, windows are not secure, and there are no physical barriers or containments systems.
T F	Employees or staff members have been assaulted, threatened, or verbally abused by clients and patients.
T F	Mediation and counseling services have **NOT** been offered to employees who have been assaulted.
T F	Alarm systems such as panic alarm buttons, silent alarms, or personal electronic alarm systems are **NOT** being used for prompt security assistance.
T F	There is no regular training provided on correct response to alarm sounding.

Table 7-4. Workplace Violence Checklist. Source: Occupational Safety and Health Administration (OSHA) (*Continued*)

T F	Alarm systems are **NOT** tested on a monthly basis to assure correct function.
T F	Security guards are **NOT** employed at the workplace.
T F	Closed circuit cameras and mirrors are **NOT** used to monitor danger areas.
T F	Metal detectors are **NOT** available or **NOT** used in the facility.
T F	Employees have **NOT** been trained to recognize and control hostile and escalating aggressive behaviors, and to manage assault behavior.
T F	Employees **CANNOT** adjust work schedules to use the "Buddy system" for visits to clients in areas where they feel threatened.
T F	Cellular phones or other communication devices are **NOT** made available to filed staff to enable them to request aid.
T F	Vehicles are **NOT** maintained on a regular basis to ensure reliability and safety.
T F	Employees work where assistance is **NOT** quickly available.

nocuous or trivial they may appear, carry the potential for violence. Therefore, prompt response to a reported threat should be a priority for management.

When appropriate, management should consult with their legal department for the best course of action to take.

Employee Relations Considerations

Knowledge of employee relations plays an integral role in the prevention and response to potentially violent workplace actions. The Employee Relations Department can:

- Help in coordinating effective responses,
- Determine the administrative options available for the removal of potentially dangerous employees;
- Determine what resources are needed depending upon the situation; and
- Ensure that disciplinary actions are taken for violent, threatening, harassing, and other disruptive behavior;
- Draft a response to an employee who raises the issue of a medical condition and/or disability as a defense against the behavior;
- Ordering and offering psychiatric examinations; and
- Provide information on appeals of disciplinary actions.

Security Measures

Maintaining a physically safe workplace is part of any good prevention program. There are a variety of security measures to help ensure safety. These include:

- Employee photo identification badges;
- On-site guard services and/or individually coded card keys for access to buildings and areas within buildings according to individual needs; and
- Assistance in registering, badging and directing visitors in larger facilities.

Workplace Security

Whether a company has its own in-house security staff, contracts with private security firms, or depends upon local law enforcement, a plan should be in place for a workplace violence prevention program involving the coordination of all these units. Planning a program for the prevention of workplace violence begins by working with and coordinating the services of the various law enforcement organizations.

Meeting with the local police department can help establish a procedure regarding law enforcement response in the event of an incident or potential incident. Once a plan has been formulated and coordinated between the company and local law enforcement, open lines of communication should exist to avoid later misunderstandings of which department—in-house, off-site or local law enforcement—is responsible when an incident occurs.

In the initial planning stage, the following can be determined by the law enforcement/security officers:

- Identify the types of situations they can address and when and how notification of an incident takes place;
- Indicate whether their officers have arrest authority;
- Identify jurisdictional restrictions;
- Identify alternative law enforcement agencies that can provide assistance;
- Identify threat assessment professionals who can assist in the protection of threatened employees;
- Explain anti-stalking laws;
- Explain how and when to obtain restraining orders;
- Suggest security measures to be taken for specific situations;
- Advise on what evidence is necessary and how it can be collected/recorded; and
- Arrange for supervisory employee briefings or training on specific workplace violence issues such as:
  - Personal safety and security measures
  - Types of incidents to report to law enforcement/security
  - Types of measures law enforcement/security may take to protect employees
  - Suggestions on how to react to an armed attacker

— Suggestions for dealing with angry customers or clients
— Suspicious packages
— Bomb threats
— Hostage situations and
— Telephone harassment and threats

The Violence Protection Program: A Basic Action Plan

One major component of any workplace violence program is prevention. In addition to program development, union involvement is an important part of a workplace violence prevention program. This section will focus on additional measures that can be taken to reduce the risk of violent behavior.

Incorporating OSHA's *Guidelines for Preventing Workplace Violence for Health Care Service Workers* into a company-specific violence prevention program includes: management commitment and employee involvement; worksite analysis; hazard prevention and control; safety and health training, and recordkeeping and program evaluation.

The primary safety officer of the company, often the facility manager, needs to assemble a "crisis management team." With top management representation and participation, the team should be comprised of key managers including internal and external safety and security personnel, medical personnel, legal counsel, human resources staff, employee assistance professionals, employee representatives and public relations support.

The specific responsibilities of the company's crisis management team are to:

1. Determine how the issue of workplace violence will be addressed in the company.

2. Develop procedures for notifying team members, internal security, local law enforcement, other employees, the general public and family members, when an emergency arises.

3. Define procedures for providing counseling to employees in terms of:
   a. What services will be provided to employees within the crucial 24-48 hours following the incident, as well as long-term services

b. Where will counseling services be provided (i.e., on site versus alternate location)

4. Develop communications strategies for addressing manager and employee concerns regarding return-to-work and pay issues. A generic press release must also be predeveloped, along with specific criteria that detail what to say and what not to say to the media. Damage control strategies that are designed to protect the company's name and reputation must also be defined.

WORKPLACE VIOLENCE AS A LIABILITY ISSUE

Because employers are required to provide a safe working environment, the question of liability takes on a special significance when a violent incident occurs. When an employee commits a violent act that injures a co-worker or third party, the employer can be held liable.

The key to determining employer liability rests with the issue of negligence. Negligence in either hiring a new employee or in retaining persons with a propensity towards violence assumes a violation of the employer's general duty to provide a safe working environment. Therefore, it is necessary to take precautions in hiring a new person, and in monitoring current employee behavior.

Hiring New Workers

The human resources department, which pre-screens job applicants before referring then to the hiring manager, can also arrange for pre-employment tests, as well as medical examinations when and where appropriate, and only in conjunction with criteria established by the spectrum of employment law (e.g., ADA).

Department and division managers along with facility managers also have an additional mechanism in place to assist them. The Introductory or Probationary Period is the most opportune time to closely scrutinize new employees and to terminate them when necessary and when it is appropriate to do so.

Retaining Employees

Retaining employees is contingent, of course, upon performance. However, other factors come into play as well. Employee behavior, adhering to policies, procedures and work practices are also taken into consideration. Any deviation from the established "norms" should be addressed quickly and aggressively.

While disciplining employees is never easy, it is essential to maintaining business and department operations, and doing so in a safe environment. Documenting an employee's poor performance and/or inappropriate behavior is the key to effective discipline. Consistency in application is the corner stone of an equitable disciplinary policy.

Limiting Exposure To Liability

In all matters related to safety management, including violence prevention, training is of paramount importance. Training that addresses the issue of discipline, recognizing "troubled employees," diffusing the hostile situation and ensuring ADA compliance, should be provided routinely to managers and supervisors.

Training that focuses on employee responsibilities to maintain a safe working environment includes:

- Policy and safe work practices and procedures.
- Employer notification of real and potential problems.
- Appropriate behavior response in potentially violent situations.
- Emergency evacuation (and self-removal) procedures.

Employee training should be conducted regularly and incorporated into the overall employee training program.

Recovery After an Incident

Planning for workplace violence protection is a necessary part of the disaster/recovery plan. It cannot, however, stop someone determined to commit violence from committing that violent act. Therefore, management should be prepared for the recovery aspect of a violent incident. Listed below are several steps management can take initially when a violent incident occurs:

- Ensure that there is a management presence in the worksite;
- Share information with employees;
- Include union leadership;

- Bring in crisis response professionals;
- Support informal debriefing;
- Support care-giving within work groups;
- Handle critical sites with care;
- Buffer those affected from post-event stresses;
- Help employees face feared places or activities; and
- Remember the healing value of work.

Ensure a Management Presence in the Worksite

When management is visible, it sends a signal to employees that they are concerned for their welfare and well-being and will answer any question/concerns that they employees do have.

Share Information with Employees

Communication between management and employees should be open—employees will have questions that need to be answered to help resolve the effects of the incident in their own minds. When information becomes available to management, management should pass that information on to employees simply and directly. "Hotlines," message boards, etc., are some effective ways to disseminate basic information. Individuals who have questions should be provided with a user-friendly system to answer those questions.

Include Union Leadership

Union representatives can help reassure employees as well as get information to them.

Use Crisis Response Professionals

Mental health professionals from either the company's Employee Assistance Program and/or the community should be brought in as soon as possible. These counselors would offer such services as debriefing, defusing and informal counseling.

Support Informal Debriefing

Employees should be provided for opportunities to talk informally even after the formal debriefing process. These informal debriefings could occur during break times.

Support Care-giving within Work Groups

Employees should not be isolated from their normal support groups at work. Employees generally will need to support and care for one another.

Handle Critical Sites with Care

The site of a violent incident will most likely be secured as a crime scene. Once the crime scene becomes "un-secured," every effort should be made to clean up the affected site without making it sterile.

Buffer Those Affected from Post-event Stresses

To reduce the pressure of the media from the vulnerable employees / family members, management should coordinate information to the media and provide timely dissemination of information.

Help Employees Face Feared Places or Activities

Having a friend or loved one to the site, or being supported by close work associates, may make returning to the site easier for the employee.

Remember the Healing Value of Work

Returning to a sense of normalcy by getting back to work can help the healing process. The return to work should also be tempered with respect for the deceased, injured and / or traumatized.

Sources

Gustin, Joseph F., *Safety Management: A Guide for Facility Managers*, New York: The Facilities Management Library. UpWord Publishing, Inc., 1996

May, Barry, *Security Management: A Guide for Facility Managers*, New York: The Facilities Management Library. UpWord Publishing, Inc., 1996.

United States Department of Agriculture, *The USDA Handbook On Workplace Violence Prevention And Response.* December, 1998.

United States Department of Labor, Bureau of Labor Statistics, *Census of Fatal Occupational Injuries 2005.*

United States Department of Labor, Occupational Safety and Health Administration, *Guidelines for Preventing Workplace Violence for Health Care and Social Service Workers*, OSHA 3148-OIR, 2004.

United States General Services Administration, Office of Personnel Management, Office Of Workforce Relations, *Dealing With Workplace Violence, A Guide for Agency Planners.* OWR-9, February, 1998.

Chapter 8

The Safety, Emergency Response & Hazard Communication Planning Program

PURPOSE

There are a number of reasons why a written safety and health plan is important.

First a well-written safety and health plan ensures compliance with the myriad of safety and health regulations mandated by OSHA and other agencies governing workplace safety and health.

Second, it provides a mechanism for incorporating other workplace mandates, such as the Americans with Disabilities Act (ADA) and worker's compensation statutes, into the company's overall hazard prevention program.

Third, it serves as the basis of the essential framework of a company's commitment to the shared safety responsibility of employers and employees.

Fourth, it allows for the equitable application and enforcement of policies, procedures, practices and rules. Equity is particularly relevant from a management perspective because it presumes consistency. Consistency in applying and enforcing rules, regulations, policies and procedures is what, in many cases, keeps managers out of trouble; and oftentimes, consistency is what keeps employers out of court.

Finally, under OSHA's current penalty structure, a company with a written safety and health plan that demonstrates commitment and involvement by all personnel receives a 25% penalty reduction based upon its good faith efforts. A 25% reduction in penalties is reason enough for facility managers to develop a written safety and health plan, while making it an "easy sell" to their employers.

ELEMENTS AND SCOPE OF THE WRITTEN SAFETY PLAN

In January 1989, OSHA issued guidelines for managing safety and health programming. These guidelines, while voluntary in nature, serve as the basis for developing a written safety and health management plan.

The guidelines are specific in identifying the elements that OSHA defines as essential to an effective safety and health management program. These elements are:

1. Management Commitment and Employee Involvement
2. Worksite Analysis
3. Hazard Prevention and Control
4. Training

The written plan should reflect the key components of effective safety and health programming that are defined by each of the four elements:

Management commitment and employee involvement are paramount to successfully implementing effective programming. OSHA, in fact, describes management commitment and employee involvement as complementary, with

1) management providing the motivation and resources of organizing and controlling activities within the organization and

2) employee involvement providing the means by which workers develop and/or express their own commitment to safety and health protection for themselves and their co-workers. In tandem, management commitment and employee involvement form the "core" of a company's occupational safety and health program, and provide facility managers with direction in compliance planning.

Management commitment and employee involvement include:

- a clearly stated worksite policy on safe and healthful work and working conditions, so that all personnel with site responsibility (and personnel at other locations with responsibility of the site) fully understand the priority and importance of safety and health protection in the organization;

- a clearly communicated goal for the safety and health program that defines objectives for meeting that goal, so that all members of the organization understand the results desired and the measures planned for achieving them;
- visible top management involvement in implementing the program, so that all employees understand that management's commitment is serious.
- provisions for employee involvement in the structure and operation of the program and in the decisions that affect their safety and health, so that they will commit their insight and energy to achieving the safety and health program's goal and objectives;
- clearly communicated and assigned responsibilities for all aspects of the program, so that managers, supervisors and employees in all parts of the organization know what performance is expected of them;
- providing adequate authority and resources to the responsible parties, so that assigned responsibilities can be met;
- holding managers, supervisors and employees accountable for meeting their responsibilities, so that essential tasks will be performed;
- reviewing program operations at least annually to evaluate their success in meeting the goals and objectives, so that deficiencies can be identified and the program and/or the objectives can be revised when they do not meet the goal of effective safety and health protection.

Worksite analysis involves various examinations of all worksite operations and job functions, so that any existing hazards and conditions can be identified. Worksite analysis includes:

- conducting comprehensive baseline worksite surveys for safety and health and periodic comprehensive update surveys;

- analyzing planned and new facility, processes, materials and equipment;
- conducting regular site safety and health inspections so that new or previously missed hazard and failures in hazard controls are identified;
- providing a reliable and reprisal-free system for employees to notify management personnel about conditions that appear hazardous and to receive timely and appropriate response (this utilizes employee insight and experience in safety and health protection and allows employee concerns to be addressed);
- investigating accidents and "near miss" incidents so that their causes can be identified and their prevention can be determined;
- analyzing injury and illness trends over time so that patterns with common causes can be identified and prevented.

Hazard prevention and control involves preventing the hazard when and where feasible, by effective job and/or worksite design and/or instituting control procedures to either eliminate or minimize hazard effects.

The procedures should address measures that:

- use engineering control techniques where feasible and appropriate;
- establish, at the earliest time, safe work practices and procedures that are clearly understood and followed by all affected parties (this includes a clearly communicated disciplinary system);
- use administrative controls, such as reducing the duration of exposure;
- provide personal protective equipment when necessary and/or when engineering and administrative controls are not feasible;
- maintain the facility and all equipment to prevent equipment breakdown;

- plan and prepare for emergencies by conducting training and emergency drills as needed to ensure the proper and safe responses to emergencies will be "second nature" for all persons involved, including the leaders who will be expected to manage and coordinate emergency response activities.
- develop a hazard communication program that alerts workers, as well as contractors and their employees to the potential risks; and
- establish a medical management program that includes on-site first aid, as well as nearby physician/emergency medical care to reduce the risk of any injury and/or illness that occurs.

Training that is job-specific must be provided to all employee levels so that they fully understand and are aware of the hazard which they may be exposed to while performing their job duties. Supervisors should be trained to:

- analyze work under their supervision to anticipate and identify potential hazards;
- maintain physical protections in their work areas;
- reinforce employee training on the nature of potential hazards in their work and on needed protective measures, through continual performance feed-back and, if necessary, through enforcement of safe work practices; and
- understand their safety and health responsibilities.

Multi-site Facilities: While corporate offices normally develop safety and health policies, procedures and practices, a company's site facilities must also develop their own "site-specific" compliance plans. These site-specific plans must address the specific operations of the site. The site's designated safety officer assumes responsibility for compliance planning.

GETTING STARTED

Prior to writing the actual plan, there are a number of preliminary steps that facility managers must take in order to assess their company's overall compliance with safety and health standards and regulations.

These preliminary steps include:

- Reviewing the company's existing safety policy;
- Analyzing current worksite practices and procedures; and
- Conducting periodic self-inspection.

Each of these steps is critical to meeting OSHA's requirement for a "safe and healthful working environment" for all employees. While the first two steps should be considered as the essential prerequisites that form the framework of the written safety and health plan, the third step—conducting periodic self-inspections—provides the basis for updating the written plan, as updating or modifying the plan becomes necessary.

Reviewing the Company's Existing Safety Policy: Policies are the official governing principles of a company. As such, policies determine the course of action that a company takes in terms of its mission and methods of operation. In short, policies provide a company's direction.

For this reason the safety policy's general statement should reflect:

- The involvement of all company personnel in safety issues; and
- The commitment of all company personnel to maintaining a safe work environment.

A good example of a general policy statement that captures the essence and spirit of shared involvement and commitment to safety is provided by OSHA in the Agency's *Handbook for Small Businesses*:

"We recognize that the responsibilities for safety and health are shared:

- *The employer accepts the responsibility for leadership of the safety and health program, for its effectiveness and improvement, and for providing the safeguards required to ensure safe conditions.*
- *Supervisors are responsible for developing the proper attitudes toward*

safety and health in themselves and in those they supervise, and for ensuring that all operations are performed with the utmost regard for the safety and health of all personnel involved, including themselves.

- *Employees are responsible for wholehearted, genuine cooperation with all aspects of the safety and health program including compliance with all rules and regulations—and for continuously practicing safety while performing their duties.*

To ensure that all persons governed by the policy understand their rights and responsibilities, the policy must clearly communicate the company's position on the issue. The policy text, therefore, should include the safety goals that are company-specific, and the methods that the company will use to achieve those goals. This is accomplished by the process shown below:

1. Define objectives for meeting the safety goals;

2. Assign specific responsibilities for all employee levels so that all persons fully understand the consequences of noncompliance;

3. Determine accountability for various employee levels so that all person fully understand the consequences of noncompliance;

4. Provide the authority, as appropriate, and the resources as necessary, to all employee levels so that they can meet their specific responsibilities; and

5. Establish a mechanism for employees to report any hazards and/or potential hazards without fear of reprisal.

Analyzing Worksite Practice and Procedures: Assessing compliance involves analyzing the current conditions at a facility and involves two major activities;

1. Undertaking a comprehensive safety and health survey of the entire facility; and

2. Evaluating existing safety and health programming so that areas of strength and weakness can be identified.

The initial survey should focus on evaluating workplace conditions with respect to safety and health regulations and generally recognized safe work practices. With such a focus, two additional benefits are derived. First, employee work habits and practices can be directly observed. Second, safety and health issues and concerns can be discussed directly with employees. OSHA's *Handbook for Small Businesses* contains an extensive self-inspection checklist that can be used to conduct the initial survey.

The second major activity in assessing an existing safety and health program is to identify areas that are working well, and to identify those areas that need improvement. Included in this activity is reviewing the four major components of a total safety and health management program. These components are:

1. Safety and Health Activities, which involves evaluating current/past activities as well as examining operations and practices, guidelines, policies and training program needs.

2. Equipment, which entails making lists of the company's major equipment, principal operations and the locations of each. Particular attention should be given to inspection schedules, maintenance activities and plan and office layouts.

3. Employee Capabilities, which involves reviewing the employment history of workers, includes hire/transfer dates, pervious and current experience and training. Particular attention should be given to newly-hired employees; employees who have handicaps and employees who have been determined to have disabilities, per ADA criteria.

4. Accident and Injury/Illness History involves tracking worker's compensation records, insurance payments first-aid cases and employee attendance records. Tracking these records can point out incidents of recurring injury/illness and point out safety areas that need to be shored-up.

The data obtained from this "fact-finding" provide very valuable information to employers. They can pinpoint any interruptions in normal operations, equipment and personnel downtime and product flaws.

Depending upon the results of this assessment, problems can be identified. Once problems are "isolated," solutions to correcting them can be determined.

Conducting Periodic Self-Inspections: The third step in assessing compliance involves periodic self-inspection. Regularly scheduled, a self-inspection can point out hazards and potential hazards that may have been missed in the initial survey. Self-inspections can also identify deficiencies in hazard control methods and procedures and, as previously noted, form the basis for modifying the written safety plan, as such modification, or updating, becomes necessary. Specifically, a self-inspection should include:

- Processing, Receiving, Shipping and Storage—equipment, job planning, layout, heights, floor loads, projection of materials, material handling and storage methods.

- Building and Grounds Conditions—floors, walls, ceilings, exits, stairs, walkways, ramps, platforms, driveways, aisles.

- Housekeeping Program—waste disposal, tools, objects, materials, leakage and spillage, cleaning methods, schedules, work areas, remote areas, and storage areas.

- Electricity—equipment, switches, breakers, fuses, switch-boxes, junctions, special fixtures, circuits, insulation, extensions, tools, motors, grounding, NEC compliance.

- Lighting—type, intensity, controls, conditions, diffusion, location, glare and shadow control.

- Heating and Ventilation—type, effectiveness, temperature, humidity, controls, natural and artificial ventilation and exhausting.

- Machinery—points of operation, flywheels, gears, shafts, pulleys, key ways, belts, couplings, sprockets, chains, frames, controls, lighting for tools and equipment, brakes, exhausting, feeding, oiling, adjusting, maintenance, lock-out, grounding, work space, location, and purchasing standards.

- Personnel—training, experience, methods, of checking machines before use, type of clothing, personal protective equipment, use of guards, tool storage, work practices, method of cleaning, oiling, or adjusting machinery.

- Hand and Power tools—purchasing standards, inspection, storage, repair, types, maintenance, grounding, use and handling.

- Chemicals—storage, handling, transportation, spills, disposals, amounts used, toxicity or other harmful effects, warning signs, supervision, training, protective clothing and equipment.

- Fire Prevention—extinguishers, alarms, sprinklers, smoking rules, exits, personnel assigned, separation of flammable materials and dangerous operations, explosion-proof fixtures in hazardous locations, waste disposal.

- Maintenance—regularity, effectiveness, training of personnel, materials and equipment used, record maintained, method of locking out machinery, general methods.

- Personal Protective Equipment—type, size, maintenance, repair, storage, assignment of responsibility, purchasing methods, standards observed, training in care and use, rules of use, method of assignment.

Easily adaptable to any company's particular situation, this outline can serve as the basis for the periodic self-inspection.

In a word, assessment means analysis. And while analyzing the worksite to ensure compliance with regulatory mandates might be its most obvious purpose, ensuring worker safety is its most immediate and compelling reason for doing so.

WRITING THE SAFETY AND HEALTH PLAN: HOW TO DO IT

The written safety and health plan does not have to be lengthy; nor does it have to be complicated. As long as the essential elements are addressed, the written safety and health plan will meet its stated purpose.

As previously discussed, the four basic elements of the effective safety and health plan, deemed essential by OSHA, include:

- Management commitment and employee involvement;
- Worksite analysis;
- Hazard prevention and control; and
- Training.

Appropriately defined, each of these essential elements speaks to a facility's total safety and health program, including emergency preparedness and hazard communications. Both of these issues must also be included within the context of the company's overall safety and health program, either as sub-sections of the plan, or as separate, "stand-alone" documents.

Review Standards: In writing the safety and health plan, it is essential to first review all applicable standards and regulations, whether promulgated by OSHA or any other agency that governs the operations of a particular industry and/or the activities of a particular facility. In reviewing the standards and regulations, facility managers must consider both general industry standards—which have an almost "universal" application—and the standards which are specific to their particular industry.

The "key" factors to identify in the standards review include:

1. Who is covered by the standard;

2. What the standard requires in terms of controls, processes and methods of operation;

3. Conditions under which those controls, processes and methods of operation are to be instituted and/or initiated; and

4. Required training for those employees whose jobs involve performing duties that are governed by the mandated controls, processes and methods of operation.

Each of these factors is critical to an effective safety and health plan because each forms the basis for compliance.

Develop Policy: Following the review of all applicable standards, the next step is to develop the safety policy. The general policy statement previously described, can serve as the basis for developing a company-specific policy that defines:

- Rights;
- Responsibilities; and
- Measures that the company will take to ensure workplace safety.

A further delineation of the rights and responsibilities of all employees, as well as the measures that the company will take to ensure workplace safety, can be broken down even further.

For example, most companies and entities already have a policy statement(s) that outlines proscribed rules of employee conduct and disciplinary procedures. The "Rules of Conduct/Discipline" normally contain:

A. Specific policy statement.
B. Applicability.
C. Guidelines for policy administration.
D. Classification of misconduct (i.e., minor vs. serious offenses).
E. Classification of disciplinary action.
F. List of rules (of conduct)-examples of misconduct.
G. Retention of disciplinary actions.

Since this particular policy statement already address these issues, it should be included in the plan itself, or at least referenced.

Flexibility: Since a safety and health program must be flexible enough to accommodate the myriad of changes that occur in the workplace, its safety and health plan must also be flexible enough to reflect those changes. As a "working" document, a company's written plan must be flexible enough to accommodate the changes to its safety and health program. The change factors that influence an entities program may occur as a result of new standards being introduced, revisions to existing standards, as well as the host of changes that occur naturally within a business setting.

In developing its essential elements for the effective safety and health program, OSHA recognized the need for flexibility. These elements provide the general direction that companies should take in developing their programs, yet provide companies with the latitude to determine how each element will be addressed.

PUTTING THE PLAN TOGETHER

As previously noted, the written safety and health plan does not have to be lengthy, nor does it have to be complicated. However, consideration must be given to those "core" components that are integral to effective safety and health programming. As such, these "core" components must also be addressed within the written plan, itself. These components are:

Regulatory standards review in which all applicable industry mandates, including emergency action plans, are reviewed to ensure compliance with defined safe work processes, practices, procedures and required employee training in same.

Policy review and development in which a review of a company/facility's existing policies is conducted to determine if the issues of safety and health are appropriately addressed within the context of business operations. Included in the policy review, is the applicability of a company/facility's "Rules of Conduct/Discipline Policy." This specific policy must be incorporated into the overall safety and health plan, so that all employees are aware of the company/facility's serious commitment to safety. Including this policy also ensure that employees at all levels will understand that they will be held accountable for meeting their safety responsibilities.

Regular worksite inspections provide facility managers with the information needed to identify, as well as isolate any problems that are either occurring, or that may occur in the workplace. Routinely conducted, worksite inspections can identify any processes, practices and procedures that carry the potential for accident and/or injury. Routinely inspecting machinery, equipment and personal protective equipment can also help facility managers in identifying potential and/or existing hazards that may have been either previously missed, or which may have surfaced since previous inspections.

Documentation, which entails all aspects of a company/facility's operations, including safety and health programming, is critical for several reasons. First, effective documentation of all safety and health initiatives provides employers with the means to track the company's safety performance and to modify and/or adapt operations, processes, procedures and training when and where necessary to meet safety/health goals and objectives. Second, effective documentation of all safety/ health initiatives, strategies and activities, demonstrates to enforcement agencies that the company/facility takes its commitment to safety and health, seriously.

Training serves as the means for providing employees with the basis for performing their assigned job tasks in a safe manner. On-going training in work processes, practices and procedures as well as in basic emergency preparedness procedures enables employees to safely meet their responsibilities, while at the same time reinforcing the fact that they will be held accountable for the safe performance of their job duties.

Sources

Gustin, Joseph F., *The Safety, Emergency Response & Hazard Communication Planning Program,* New York: UpWord Publishing, 1996.

Chapter 9

Preparing for The OSHA Inspection

In most cases, without advance notice, OSHA or the applicable state agency may visit to conduct a safety inspection. Non-compliance of OSHA regulations can result in expensive penalties. Advance planning is the key to successfully managing the inspection while the most careful preparation will not guarantee a violation-free inspection; it can minimize the impact that any ensuing litigation may have on the horizon.

THE OSHA INSPECTION

The Occupational Safety and Health Administration (OSHA) is authorized by the OSH Act of 1970 to conduct workplace inspections. These inspections are carried out to determine whether employers are complying with Agency standards, as well as to enforce Section 5(s)(1) of the OSH Act, commonly known as the General Duty Clause.

Workplace inspections are performed by Federal OSHA safety and health compliance officers for those states under federal jurisdiction. State Plan States' compliance officers perform worksite inspections in those states that operate under a federal OSHA-approved safety and health program.

Workplace inspections and investigations are conducted by OSHA compliance safety and health officers who are safety and health professionals trained in the disciplines of safety and industrial hygiene.

Inspections are always conducted without advance notice. There are, however, special circumstances under which OSHA may give notice to the employer, but such a notice will normally be less than 24 hours. These circumstances include the following:

- Imminent danger situations that require correction as soon as possible;

- Accident investigations where the employer has notified the agency of a fatality or catastrophe;
- Inspections that must take place after regular business hours or that require special preparation;
- Cases where notice is required to ensure that the employer and employee representative or other personnel will be present;
- Cases where an inspection must be delayed for more than 5 working days when there is good cause; and
- Situations in which the OSHA Area Director determines that advance notice would produce a more thorough or effective inspection.

Employers who receive advance notice of an inspection must inform their employees' representative or arrange for OSHA to do so. If an employer refuses to admit an OSHA compliance officer or if an employer attempts to interfere with the inspection, the *Act* permits appropriate legal action, such as obtaining a warrant to inspect.

INSPECTION PRIORITIES

Since not all 111 million plus workplaces covered by the Act can be inspected immediately, OSHA has established a system of inspection priorities. They are as follows:

- Imminent danger
- Catastrophes and fatal accidents
- Complaints and referrals
- Programmed inspections
- Follow-up inspections

Imminent Danger

The top priority is, as should be, situations which pose an imminent danger. An imminent danger exists when there is reasonable certainty that a danger exists that can cause immediate death or serious physical harm.

An imminent danger also exits when there is reasonable certainty that the danger can be eliminated through normal enforcement procedures.

If a compliance officer finds an imminent danger situation, the employer will be asked to voluntarily abate the hazard and remove endangered employees from exposure. Should the employer fail to do this, OSHA, through the regional solicitor, may apply to the Federal District Court for an injunction prohibiting further work as long as unsafe conditions exist.

Catastrophes and Fatal Accidents

Second priority is given to fatalities and accidents resulting in a death or hospitalization of three or more employees. In all such situations, the employer must report the incident(s) to OSHA within eight (8) hours of the occurrence or within eight hours of learning of the occurrence. Inspections of catastrophes and fatal accidents are conducted to determine the cause(s) of these accidents and whether existing OSHA standards were violated.

Complaints and Referrals

Third priority is given to formal employee complaints of alleged standards violations, or of unsafe or unhealthy working conditions.

Under the OSH Act, employees have the right to request an OSHA inspection when they believe that there is an imminent danger from a hazard or when there is a violation of an OSHA standard that threatens physical harm. OSHA will maintain confidentiality if requested, inform the employee of any action it takes regarding complaints, and, if requested, hold an informal review of any decision not to inspect.

Programmed Inspections

Fourth priority is given to programmed inspections. These inspections are aimed at specific high-hazard industries, workplaces, occupations, or health substances, as well as industries identified in OSHA's current inspection procedures.

Criteria for selecting a particular industry for programmed inspection include injury accident rates, previous citation history, employee exposure to toxic substances and random selection.

Additionally, OSHA may develop special emphasis programs which may be local, regional, or national in scope, depending on the distribution of the workplaces involved.

Comprehensive safety inspections are also conducted in those manufacturing companies with lost-workday injury rates at or above the Bureau of Labor Statistics (BLS) national rate for manufacturing currently in use by OSHA. States with their own occupational safety and health programs may use somewhat different systems to identify industries for inspection.

Follow-up Inspections

Last in priority is the follow-up inspection, which is conducted to determine if previously cited violations have been corrected. If the employer has failed to abate a violation, the employer is subject to "Failure to Abate" citations. Such citations can result in daily penalties until the employer corrects the violations(s).

How Does a Compliance Officer Prepare for the Inspection?

A compliance officer represents the agency and is expected to demonstrate knowledge and expertise in the safety and health field in a courteous and professional manner. Before the inspection, the compliance officer will become familiar with as many relevant facts as possible about the workplace, such as its inspection history, the nature of the business, and the particular standards that might apply. This preparation provides the compliance officer with knowledge of the potential hazards and industrial processes that may be encountered and aids in selecting appropriate personal protective equipment for use against these hazards during the inspection.

THE INSPECTION PROCESS

Inspector's Credentials

When the OSHA compliance officer arrives at the establishment, the officer will display official credentials and ask to meet an appropriate employer representative. Employers should always ask to see the compliance officer's credentials.

Employers may verify the OSHA federal or state compliance officer credentials by calling the nearest federal or state OSHA office. Compliance officers may not collect a penalty at the time of the inspection or promote the sale of a product or service at any time. Anyone who attempts to do so is impersonating a government inspector and the employer should contact the FBI or local law enforcement officials *immediately*.

Opening Conference

In the opening conference, the compliance officer explains how the establishment was selected and what the likely scope of the inspection will be. The compliance officer also will ascertain whether an OSHA-funded consultation visit is in progress or whether the facility is pursuing or has received an inspection exemption through the consultation program. If so, the inspection may be limited or terminated.

The compliance officer explains the purpose of the visit, the scope of the inspection, and the standards that apply. The compliance officer gives the employer information on how to get a copy of applicable safety and health standards as well as a copy of any employee complaint that may be involved (with the employee's name deleted, if the employee requests anonymity).

The compliance officer asks the employer to select an employer representative to accompany the compliance officer during the inspection.

The compliance officer also gives an authorized employee representative the opportunity to attend the opening conference and accompany the compliance officer during the inspection. If a recognized bargaining agent represents the employees, the agent ordinarily will designate the employee representative to accompany the compliance officer. Similarly, if there is a plant safety committee, the employee members of that committee will designate the employee representative (in the absence of a recognized bargaining agent). Where neither employee group exists, the employees themselves may select an employee representative, or the compliance officer may determine if any employee suitably represents the interest of other employees.

The *Act* does not require an employee representative for each inspection. Where there is no authorized employee representative, however, the compliance officer must consult with a reasonable number of employees concerning safety and health matters in the workplace.

Walkthrough

After the opening conference, the compliance officer and accompanying representatives proceed through the establishment to inspect work areas for safety and health hazards.

The compliance officer determines the route and duration of the inspection. While talking with employees, the compliance officer makes every effort to minimize any work interruptions. The compliance officer observes safety and health conditions and practices; consults with employees privately, if necessary; takes photos, videotapes, and instrument readings; examines records; collects air samples; measures noise levels; surveys existing engineering controls; and monitors employee exposure to toxic fumes, gases, and dusts.

An inspection tour may cover part or all of an establishment, even if the inspection resulted from a specific complaint, fatality, or catastrophe. If the compliance officer finds a violation in open view, he or she may ask permission to expand the inspection.

The compliance officer keeps all trade secrets observed confidential.

The compliance officer consults employees during the inspection tour. The officer may stop and question workers, in private, about safety and health conditions and practices in their workplaces. Each employee is protected under the *Act* from discrimination by the employer for exercising their safety and health rights.

OSHA places special importance on posting and recordkeeping requirements. The compliance officer will inspect records of deaths, injuries, and illnesses that the employer is required to keep. The officer will also check to see that a copy of the totals from the last page of OSHA Form Number 300 are posted as required and that the OSHA workplace poster, which explains employees' safety and health rights, is prominently displayed (see Figure 9-1). Where records of employee exposure to toxic substances and harmful physical agents are required, the compliance officer will examine them for compliance with the recordkeeping requirements.

The compliance officer also requests a copy of the employer's Hazard Communication Program. Under OSHA's Hazard Communication Standard, employers must establish a written, comprehensive communication program that includes provisions for container labeling, material safety data sheets, and an employee training program. The program must contain a list of the hazardous chemicals in each work area and the means the employer will use to inform employees of the hazards associated with these chemicals.

During the course of the inspection, the compliance officer will point out to the employer any unsafe or unhealthful working conditions observed. At the same time, the compliance officer will discuss possible corrective action if the employer so desires.

Some apparent violations detected by the compliance officer can be corrected immediately. When the employer corrects them on the spot, the compliance officer records such corrections to help in judging the

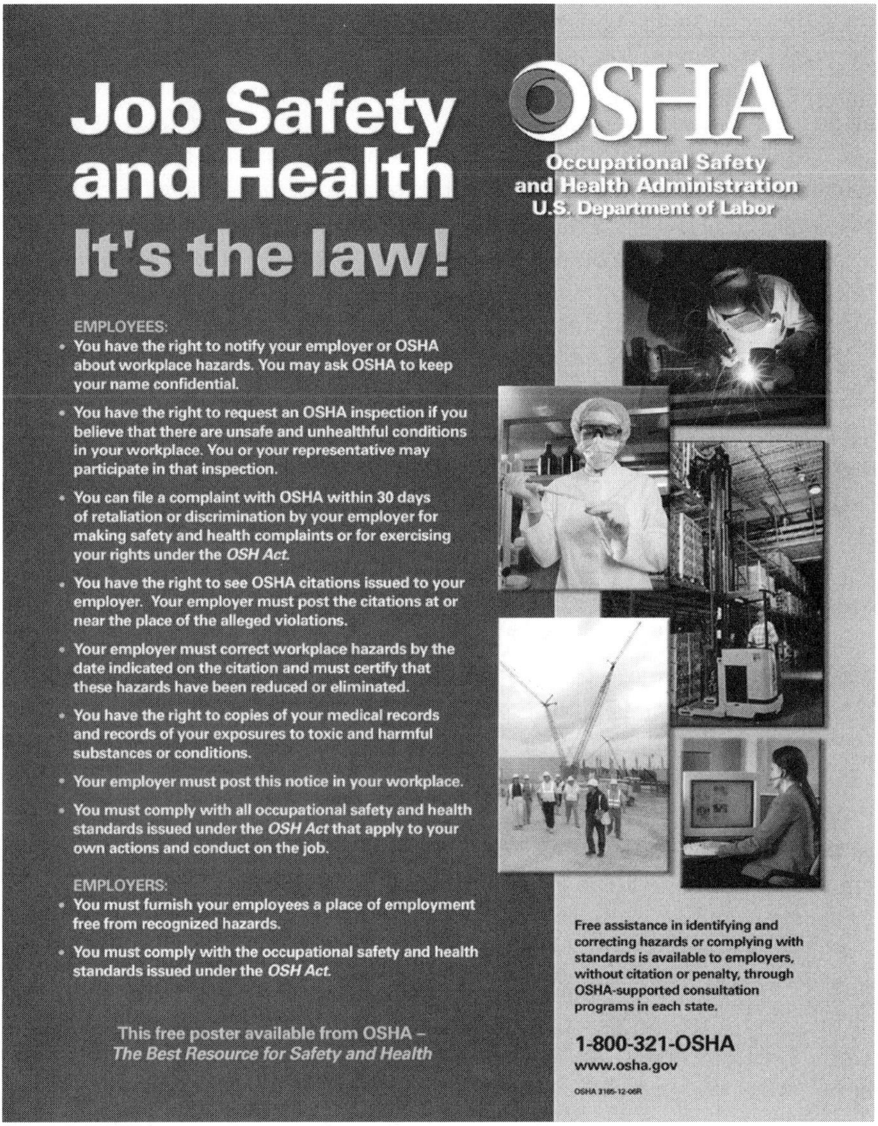

Source: OSHA

Figure 9-1.

employer's good faith in compliance. Although corrected, the apparent violations will serve as the basis for a citation and, if appropriate, a notice of proposed penalty. OSHA may reduce the penalties for some types of violations if they are corrected immediately.

Closing Conference

At the conclusion of the inspection, the compliance officer conducts a closing conference with the employer, employees, and/or the employees' representative.

The compliance officer gives the employer and all other parties involved a copy of *Employer Rights and Responsibilities Following an OSHA Inspection* (OSHA 3000) for their review and discussion.

The compliance officer discusses with the employer all unsafe or unhealthful conditions observed during the inspection and indicates all apparent violations for which a citation and a proposed penalty may be issued or recommended. The compliance officer will not indicate any specific proposed penalties but will inform the employer of appeal rights.

During the closing conference, the employer may wish to produce records to show compliance efforts and provide information that can help OSHA determine how much time may be needed to abate an alleged violation.

When appropriate, the compliance officer may hold more than one closing conference. This is usually necessary when the inspection includes an evaluation of health hazards, after a review of additional laboratory reports, or after the compliance officer obtains additional factual evidence while concluding an accident investigation.

The compliance officer explains that OSHA area offices are full-service resource centers that inform the public of OSHA activities and programs. This includes information on new or revised standards, the status of proposed standards, comment periods, or public hearings. Additionally, area offices provide technical experts and materials and refer callers to other agencies and professional organizations as appropriate. The area offices promote effective safety and health programs through Voluntary Protection Programs (VPP) and provide information about study courses offered at the OSHA Training Institute or its satellite locations nationwide.

If an employee representative does not participate in either the opening or the closing conference held with the employer, the compliance officer holds a separate discussion with the employee representative, if requested, to discuss matters of direct interest to employees.

THE INSPECTION RESULTS

After the compliance officer reports findings, the Area Director determines whether citations and/or propose penalties will be issued.

Citations

Citations inform the employer and employees of the regulations and standards alleged to have been violated and of the proposed length of time set to correct alleged hazards. The employer will receive citations and notices of proposed penalties by certified mail. The employer must post a copy of each citation at or near the place a violation occurred for 3 days or until the violation is abated, whichever is longer.

Penalties

The following violations and the concomitant penalties follow:

1. *Other-Than-Serious Violation.* This violation has a direct relationship to job safety and health, but probably would not cause death or serious physical harm. OSHA may assess a penalty from $0 to $1,000 for each violation. The agency may adjust a penalty for the *other-than-serious* violation downward, by as much as 95 percent, depending on the employer's good faith (demonstrated efforts to comply with the Act), history of previous violations, and size of business.

2. *Serious Violation.* This violation indicates that there is a substantial probability that death or serious physical harm could result. OSHA assesses the penalty for a serious violation from $1,500 to $7,000 depending on the gravity of the violation. OSHA may adjust a penalty for a serious violation downward, based on the employer's good faith, history of previous violations, and size of business.

3. *Willful Violation.* A willful violation is one in which the employer intentionally and knowingly commits, or a situation in which the employer was aware that a hazardous condition existed and made no reasonable effort to eliminate it.. OSHA may propose penalties of up to $70,000 for each willful violation. The minimum willful penalty is $5,000.

   An employer who is convicted in a criminal proceeding of a willful violation of a standard that has resulted in the death of an employee may be fined up to $250,000 (or $500,000 if the employer is a corporation) or imprisoned up to 6 months, or both. A second conviction doubles the possible term of imprisonment.

4. *Repeated Violation.* A repeated violation is one in which any standard, regulation, rule, or order where, upon reinspection, a sub-

stantially similar violation is found. Violations can bring a fine or up to $70,000 for each such violation within the previous 3 years. To calculate repeated violations, OSHA adjusts the initial penalty for the size and then multiplies by a factor of 2, 5, or 10 depending on the size of the business.

5. *Failure-to-Abate*. Failure to correct a prior violation may bring a civil penalty of up to $7,000 for each day that the violation continues beyond the prescribed abatement date.

Additional violations for which OSHA may issue citations and proposed penalties are as follows:

- Falsifying records, reports, or applications can, upon conviction, bring a criminal fine of $10,000 or up to 6 months in jail, or both.

- Violating posting requirements may bring a civil penalty of $7,000.

- Assaulting a compliance officer or otherwise resisting, opposing, intimidating, or interfering with a compliance officer in the performance of the officer's duties is a criminal offense and is subject to a fine of not more than $5,000 and imprisonment for not more than 3 years.

Citations and penalty procedures may differ somewhat in states with their own occupational safety and health programs.

THE APPEALS PROCESS

Appeals by Employees

If an employee complaint initiates an inspection, the employee or authorized employee representative may request an informal review of any decision not to issue a citation.

Employees may not contest citations, amendments to citations, proposed penalties, or lack of penalties. They may, however, contest the time allowed for abatement of a hazardous condition. They also may contest an employer's "Petition for Modification of Abatement," which requests an extension of the proposed abatement period. Employees must contest the petition within 10 working days of its posting or within 10 working days after an authorized employee representative receives a copy.

Employees may request an informal conference with OSHA to discuss any issues raised by an inspection, citation, notice of proposed penalty, or employer's notice of intention to contest.

Appeals by Employers

Within 15 working days of receiving a citation, an employer who wishes to contest must submit a written objection to OSHA. The OSHA Area Director forwards the objection to the Occupational Safety and Health Review Commission (OSHRC), which operates independently of OSHA.

When issued a citation and notice of proposed penalty, an employer may request an informal meeting with OSHA's Area Director to discuss the case. OSHA encourages employers to have informal conferences with the Area Director if the employer has issues arising from the inspection that the employer wishes to discuss or provide additional information. The Area Director is authorized to enter into settlement agreements that revise citations and penalties to avoid prolonged legal disputes and result in speedier hazard abatement. (Alleged violations contested before OSHRC do not need to be corrected until the contest is ruled upon by OSHRC.)

Petition for Modification of Abatement

After receiving a citation, the employer must correct the cited hazard by the abatement date unless the citation or abatement date is contested. Factors beyond the employer's control, however, may prevent the completion of corrections by that date. In such a situation, the employer who has made a good-faith effort to comply may file a petition to modify the abatement date.

The written petition must specify the steps taken to achieve compliance, the additional time needed to comply, the reasons additional time is needed, and interim steps taken to safeguard employees against the cited hazard during the intervening period. The employer must certify a copy of the petition has been posted in a conspicuous place at or near each place where a violation occurred and that the employee representative received a copy of the petition.

Notice of Contest

If the employer decides to contest the citation, the abatement period, or the proposed penalty, the employer has 15 working days from the time the citation and proposed penalty are received to notify the OSHA Area Director. This notification must be done in writing. Failure to do so results in the citation and proposed penalty becoming a final order of the OSHRC without further appeal. An orally expressed disagreement will not suffice. This written notification is called a "Notice of Contest."

Although there is no specific format for the Notice of Contest, it must clearly identify the employer's basis for filing—the citation, notice of proposed penalty, abatement period, or notification of failure to correct violations.

The employer must give a copy of the Notice of Contest to the employees' authorized representative. If any affected employees are not represented by a recognized bargaining agent, the employer must post a copy of the notice in a prominent location in the workplace or give it personally to each unrepresented employee.

Review Procedure

If the employer files a written Notice of Contest within the required 15 working days, the OSHA Area Director forwards the case to OSHRC. The commission is an independent agency not associated with OSHA or the Department of Labor. The commission assigns the case to an administrative law judge.

OSHRC may schedule a hearing at a public place near the employer's workplace. The employer and the employee have the right to participate in the hearing. The OSHRC does not require them to be represented by attorneys.

Once the administrative law judge has ruled, any party to the case may request a further review by OSHRC. Any of the three OSHRC commissioners also may, at their own motion, bring a case before the commission for review. Employers and other parties may appeal commission rulings to the appropriate U.S. Court of Appeals.

Sources

Gustin, Joseph F., *Safety Management, A Guide for Facility Managers*, UpWord Publishing, New York, 1996.

U.S. Department of Labor, Occupational Safety and Health Administration, *What to Expect During OSHA's Visit*, 1997 updated February, 2005.

U.S. Department of Labor, Occupational Safety and Health Administration, *All About OHSA*, OSHA 3302-06-2005.

U.S. Department of Labor, Occupational Safety and Health Administration, *It's the Law!* OSHA 3165-12 O6R, 2007.

Chapter 10

Keeping Accurate Records

Accurate recordkeeping serves several purposes. First, maintaining accurate injury and illness records meets compliance obligations. It can also be used as a management tool in conjunction with a company's various cost-containment strategies, including workers' compensation and other loss prevention programs. Finally, injury and illness records can be used as a diagnostic tool to identify and correct problems.

INTRODUCTION

The Occupational Safety and Health Act of 1970 and 29 CFR Part 1904 requires employers to prepare and maintain records of all occupational injuries and illnesses. Under the Act, the Secretary of Labor is responsible for collecting, compiling and analyzing statistics of work-related injuries and illnesses. The Bureau of Labor Statistics (BLS) administers the recordkeeping and reporting system. In most states, a state agency cooperates with the BLS in administering these programs.

Injury and illness records provide a basis for a statistical program which produces data that measure OSHA's efforts. The records serve two other purposes:

- They are helpful to employers and employees in identifying factors that cause workplace injuries and illnesses.

- They are designed to assist safety and health compliance officers in making OSHA inspections.

OSHA also believes that recordkeeping is important for the employer's safety and health efforts for the following reasons:

- Keeping track of work-related injuries and illnesses can help the prevention of them in the future.

- Using injury and illness data helps identify problem areas. The more that is known, the better the identification and correction of hazardous workplace conditions.

- The company safety and health programs can be better administered with accurate records.

- As employee awareness about injures, illnesses, and hazards in the workplace improves, workers are more likely to follow safe work practices and report workplace hazards.

Armed with the data from a particular company/industry, OSHA compliance officers can rely on the data to help them properly identify and focus on injuries and illnesses in a particular area. The agency also asks about 80,000 establishments each year to report the data directly to OSHA, which uses the information as part of its site-specific inspection targeting program. The Bureau of Labor Statistics (BLS) also uses injury and illness records as the source data for the Annual Survey of Occupational Injuries and Illnesses that shows safety and health trends nationwide and industry wide.

RECORDKEEPING AGENCIES

The government agencies involved in OSHA recordkeeping include OSHA itself, the BLS and state agencies.

The Occupational Safety and Health Administration

OSHA is part of the U.S. Department of Labor. The Agency is responsible for developing, implementing and enforcing safety and health standards and regulations.

Bureau of Labor Statistics

The Bureau of Labor Statistics is also part of the U.S. Department of Labor. It is responsible for both administering and maintaining the OSHA recordkeeping system and analyzing work injury and illness statistics.

State Agencies

State agencies cooperate with the BLS in administering the OSHA recordkeeping and reporting programs. States that have their own safety and health laws may impose additional obligations upon employers.

RECORDKEEPING REQUIREMENTS

OSHA's recordkeeping requirements apply to all private sector employers in all states, the District of Columbia, Puerto Rico, the Virgin Islands, American Samoa, Guam and the Trust Territories of the Pacific Islands.

Employers Required To Keep Records

Employers in the following industries are required by law to keep OSHA records:

- Agriculture, Forestry and Fishing
- Oil and Gas Extraction
- Construction
- Manufacturing
- Transportation and Public Utilities
- Wholesale Trade
- Building Materials and Garden Supplies
- General Merchandise and Food Stores
- Hotels and Other Lodging places
- Repair Services
- Amusement and Recreation Services
- Health Services

Multi-establishment employers in any of these industries with a combined employment of 11 or more employees must keep records for each individual establishment.

Exempted Employers

Employers who normally do not have to keep OSHA records include all employers with no more than 10 full- or part-time employees at any one time in the previous calendar year and employers in certain retail trade, finance, insurance, real estate and service industries including:

- Automotive Dealers and Gasoline Service Stations
- Apparel and Accessory Stores
- Furniture, Home Furnishings and Equipment Stores
- Eating and Drinking Places
- Miscellaneous Retail
- Banking
- Credit Agencies Other than Banks
- Security, Commodity Brokers and Services
- Insurance
- Real Estate

Employers are not required to keep OSHA injury and illness records for any establishment classified in the specific Standard Industrial Classification (SIC) codes, unless they are asked in writing to do so by OSHA, the Bureau of Labor Statistics (BLS), or a state agency that operates under the authority of OSHA or the BLS. Table 10-1 lists the exempted employers. However, all employers, including those that are partially exempted by reason of company size or industry classification, must report to OSHA any workplace incident that results in a fatality or the hospitalization of three or more employees (§1904.39).

A rotating sample of these employers, however, is selected each year to keep records and to participate in a mandatory statistical survey of occupational illnesses and injuries.

The participation of these normally exempted employers is necessary for producing national estimates of occupational injures and illnesses for all exempt and non-exempt private sector employers. If regularly exempted employers are selected to maintain records and participate in the Annual Survey of Occupational Injuries and Illnesses, they will be notified in advance.

While recordkeeping requirements are reduced for these normally exempted employers, they must still comply with OSHA standards, including displaying the OSHA poster and reporting fatalities and/or multiple hospitalizations within eight hours of occurrence.

Permanently Exempted Employers

Employers and individuals that are permanently exempted from keeping OSHA injury and illness records include self-employed people, partners with no employees, employers of domestics in private residences for the purposes of housekeeping and/or childcare, and employ-

Table 10-1. Exempted Employers

SIC code	Industry description	SIC code	Industry description
525	Hardware Stores	725	Shoe Repair & Shoeshine Parlors
542	Meat & Fish Markets	726	Funeral Service & Crematories
544	Candy, Nut, Confectionery Stores	729	Misc. Personal Services
545	Dairy Products Stores	731	Advertising Services
546	Retail Bakeries	732	Credit Reporting, Collection Services
549	Miscellaneous Food Stores	733	Mailing, Reproduction, Stenographic Services
551	New & Used Car Dealers	737	Computer, Data Processing Services
552	Used Car Dealers	738	Misc.Business Services
554	Gasoline Service Stations	764	Reupholstery & Furniture Repair
557	Motorcycle Dealers	78	Motion Picture
56	Apparel & Accessory Stores	791	Dance Studios, Schools, Halls
573	Radio, Television, & Computer Stores	792	Producers, Orchestras, Entertainers
58	Eating & Drinking Places	793	Bowling Centers
591	Drug Stores & Proprietary Stores	801	Offices & Clinics Of Medical Doctors
592	Liquor Stores	802	Offices &Clinics Of Dentists
594	Misc.Shopping Goods Stores	803	Offices Of Osteopathic
599	Retail Stores, Not Elsewhere Classified	804	Offices Of Other Health Practitioners
60	Depository Institutions (banks & savings institutions)	807	Medical & Dental Laboratories
61	Non-depository Institutions (Banks & Savings Institutions)	809	Health & Allied Services, Not Elsewhere Classified
62	Security & Commodity Brokers	81	Legal Services
63	Insurance Carriers	82	Educational Services (schools, colleges, universities, libraries)
64	Insurance Agents, Brokers & Services	832	Individual & Family Services
653	Real Estate Agents & Managers	835	Child Day Care Services
654	Title Abstract Offices	839	Social Services, Not Elsewhere Classified
67	Holding & Other Investment Offices	841	Museums & Art Galleries
722	Photographic Studios, Portrait	86	Membership Organizations
723	Beauty Shops	87	Engineering, Accounting, Research, Management, & Related Services
724	Barber Shops	899	Services, not elsewhere classified

Source: OSHA

ers engaged in religious activities that concern the conduct of religious services.

In the case of the latter, records must be kept of injuries or illnesses when employees are engaged in secular activities. Private hospitals and/ or commercial establishments owned or operated by religious organizations are required to keep records.

State and Local Government

State and local government agencies are normally exempted from OSHA recordkeeping requirements. However, some states require that their agencies and local governments keep injury and illness records in accordance with state regulations.

Other Federal Regulations

Employers who are subject to other federal safety and health recordkeeping requirements are not exempted from OSHA requirements. Records used to comply with other Federal regulations may be used to satisfy OSHA requirements, however, if those forms are equivalent to the OSHA No. 300 log and summary and the OSHA No. 101 supplementary record.

RECORDKEEPING FORMS AND RECORDING PROCEDURES

Since January 1, 2002, employers have a new system for tracking workplace injuries and illnesses. OSHA's new recordkeeping log is easier to understand and to use. It is written in plain language that uses a question and answer format. The revised recordkeeping rule answers questions about recording occupational injuries and illnesses. The revised recordkeeping rule also explains how to classify particular cases.

The New Rule

This revised rule includes several key highlights. Most notably, provisions are made for:

- Computerizing injury and illness records; as well as
- Updating the three recordkeeping forms: Form 300, Form 301 and Form 300A.

Note that OSHA Form 300 (Log of Work-Related Injuries and Illnesses) is simplified and is reformatted to fit legal size paper (see Figure 10-1). Also, OSHA Form 301 (Injury and Illness Incident Report) includes more data about how the injury or illness occurred (see Figure 10-2). The OSHA Form 300A (Summary of Work-Related Injuries and Illnesses) is a separate form created to make it easier to calculate incidence rates (see Figure 10-3).

Additionally, the new rule:

- Continues to exempt smaller employers (employers with 10 or fewer employees) from most requirements;
- Changes the exemptions for employers in service and retail industries;
- Clarifies the definition of work relationship, limits the recording of pre-existing cases and adds new exceptions for some categories of injury and illness;
- Includes new definitions of medical treatment, first aid, and restricted work to simplify recording decisions;
- Eliminates different criteria for recording work-related injuries and work-related illnesses; one set of criteria will be used for both;
- Changes the recording of needlestick injuries and tuberculosis; Simplifies the counting of days away from work, restricted days and job transfer;
- Improves employee involvement and provides employees and their representatives with access to the information; and
- Finally, the new rule protects privacy for injured and ill workers.

Cases that need to be recorded are easier to determine because the definitions are made simpler and clearer. The annual summary of workplace injuries and illnesses must be posted for a longer period of time. The longer posting improves employee access to information; and ensures an increase in their involvement and participation as employees learn how to report workplace injuries and illnesses.

OSHA's Form 300 (Rev. 01/2004)

Log of Work-Related Injuries and Illnesses

Attention: This form contains information relating to employee health and must be used in a manner that protects the confidentiality of employees to the extent possible while the information is being used for occupational safety and health purposes.

Year 20__ __

U.S. Department of Labor
Occupational Safety and Health Administration

Form approved OMB no. 1218-0176

You must record information about every work-related death and about every work-related injury or illness that involves loss of consciousness, restricted work activity or job transfer, days away from work, or medical treatment beyond first aid. You must also record significant work-related injuries and illnesses that are diagnosed by a physician or licensed health care professional. You must also record work-related injuries and illnesses that meet any of the specific recording criteria listed in 29 CFR Part 1904.8 through 1904.12. Feel free to use two lines for a single case if you need to. You must complete an Injury and Illness Incident Report (OSHA Form 301) or equivalent form for each injury or illness recorded on this form. If you're not sure whether a case is recordable, call your local OSHA office for help.

Establishment name ______________

City ______________ State ______________

Identify the person			Describe the case			Classify the case: CHECK ONLY ONE box for each case based on the most serious outcome for that case:				Enter the number of days the injured or ill worker was:		Check the "Injury" column or choose one type of illness: (M)					
(A) Case no.	(B) Employee's name	(C) Job title (*e.g., Welder*)	(D) Date of injury or onset of illness	(E) Where the event occurred (*e.g., Loading dock north end*)	(F) Describe injury or illness, parts of body affected, and object/substance that directly injured or made person ill (*e.g., Second degree burns on right forearm from acetylene torch*)	Death (G)	Days away from work (H)	Remained at Work: Job transfer or restriction (I)	Remained at Work: Other record-able cases (J)	Away from work (K)	On job transfer or restriction (L)	Injury (1)	Skin disorder (2)	Respiratory condition (3)	Poisoning (4)	Hearing loss (5)	All other illnesses (6)
___	___	___	__/__ month/day	___	___	☐	☐	☐	☐	___ days	___ days	☐	☐	☐	☐	☐	☐
___	___	___	__/__ month/day	___	___	☐	☐	☐	☐	___ days	___ days	☐	☐	☐	☐	☐	☐
___	___	___	__/__ month/day	___	___	☐	☐	☐	☐	___ days	___ days	☐	☐	☐	☐	☐	☐
___	___	___	__/__ month/day	___	___	☐	☐	☐	☐	___ days	___ days	☐	☐	☐	☐	☐	☐
___	___	___	__/__ month/day	___	___	☐	☐	☐	☐	___ days	___ days	☐	☐	☐	☐	☐	☐
___	___	___	__/__ month/day	___	___	☐	☐	☐	☐	___ days	___ days	☐	☐	☐	☐	☐	☐
___	___	___	__/__ month/day	___	___	☐	☐	☐	☐	___ days	___ days	☐	☐	☐	☐	☐	☐
___	___	___	__/__ month/day	___	___	☐	☐	☐	☐	___ days	___ days	☐	☐	☐	☐	☐	☐
___	___	___	__/__ month/day	___	___	☐	☐	☐	☐	___ days	___ days	☐	☐	☐	☐	☐	☐
___	___	___	__/__ month/day	___	___	☐	☐	☐	☐	___ days	___ days	☐	☐	☐	☐	☐	☐
___	___	___	__/__ month/day	___	___	☐	☐	☐	☐	___ days	___ days	☐	☐	☐	☐	☐	☐
___	___	___	__/__ month/day	___	___	☐	☐	☐	☐	___ days	___ days	☐	☐	☐	☐	☐	☐
___	___	___	__/__ month/day	___	___	☐	☐	☐	☐	___ days	___ days	☐	☐	☐	☐	☐	☐
					Page totals ➤	___	___	___	___	___	___	___ Injury (1)	___ Skin disorder (2)	___ Respiratory condition (3)	___ Poisoning (4)	___ Hearing loss (5)	___ All other illnesses (6)

Be sure to transfer these totals to the Summary page (Form 300A) before you post it.

Public reporting burden for this collection of information is estimated to average 14 minutes per response, including time to review the instructions, search and gather the data needed, and complete and review the collection of information. Persons are not required to respond to the collection of information unless it displays a currently valid OMB control number. If you have any comments about these estimates or any other aspects of this data collection, contact: US Department of Labor, OSHA Office of Statistical Analysis, Room N-3644, 200 Constitution Avenue, NW, Washington, DC 20210. Do not send the completed forms to this office.

Page ___ of ___

Figure 10-1. OSHA's Form 300.

OSHA's Form 301

Injury and Illness Incident Report

Attention: This form contains information relating to employee health and must be used in a manner that protects the confidentiality of employees to the extent possible while the information is being used for occupational safety and health purposes.

U.S. Department of Labor
Occupational Safety and Health Administration

Form approved OMB no. 1218-0176

This *Injury and Illness Incident Report* is one of the first forms you must fill out when a recordable work-related injury or illness has occurred. Together with the *Log of Work-Related Injuries and Illnesses* and the accompanying *Summary*, these forms help the employer and OSHA develop a picture of the extent and severity of work-related incidents.

Within 7 calendar days after you receive information that a recordable work-related injury or illness has occurred, you must fill out this form or an equivalent. Some state workers' compensation, insurance, or other reports may be acceptable substitutes. To be considered an equivalent form, any substitute must contain all the information asked for on this form.

According to Public Law 91-596 and 29 CFR 1904, OSHA's recordkeeping rule, you must keep this form on file for 5 years following the year to which it pertains.

If you need additional copies of this form, you may photocopy and use as many as you need.

Completed by __________

Title __________

Phone (____)____-____ Date ___/___/___

Information about the employee

1) Full name __________

2) Street __________

City ______ State ____ ZIP ____

3) Date of birth ___/___/___

4) Date hired ___/___/___

5) ☐ Male
☐ Female

Information about the physician or other health care professional

6) Name of physician or other health care professional __________

7) If treatment was given away from the worksite, where was it given?

Facility __________

Street __________

City ______ State ____ ZIP ____

8) Was employee treated in an emergency room?
☐ Yes
☐ No

9) Was employee hospitalized overnight as an in-patient?
☐ Yes
☐ No

Information about the case

10) Case number from the *Log* ______ *(Transfer the case number from the Log after you record the case.)*

11) Date of injury or illness ___/___/___

12) Time employee began work ______ AM / PM

13) Time of event ______ AM / PM ☐ Check if time cannot be determined

14) ***What was the employee doing just before the incident occurred?*** Describe the activity, as well as the tools, equipment, or material the employee was using. Be specific. *Examples:* "climbing a ladder while carrying roofing materials"; "spraying chlorine from hand sprayer"; "daily computer key-entry."

15) ***What happened?*** Tell us how the injury occurred. *Examples:* "When ladder slipped on wet floor, worker fell 20 feet"; "Worker was sprayed with chlorine when gasket broke during replacement"; "Worker developed soreness in wrist over time."

16) ***What was the injury or illness?*** Tell us the part of the body that was affected and how it was affected; be more specific than "hurt," "pain," or sore." *Examples:* "strained back"; "chemical burn, hand"; "carpal tunnel syndrome."

17) ***What object or substance directly harmed the employee?*** *Examples:* "concrete floor"; "chlorine"; "radial arm saw." *If this question does not apply to the incident, leave it blank.*

18) ***If the employee died, when did death occur?*** Date of death ___/___/___

Public reporting burden for this collection of information is estimated to average 22 minutes per response, including time for reviewing instructions, searching existing data sources, gathering and maintaining the data needed, and completing and reviewing the collection of information. Persons are not required to respond to the collection of information unless it displays a current valid OMB control number. If you have any comments about this estimate or any other aspects of this data collection, including suggestions for reducing this burden, contact: US Department of Labor, OSHA Office of Statistical Analysis, Room N-3644, 200 Constitution Avenue, NW, Washington, DC 20210. Do not send the completed forms to this office.

Figure 10-2. OSHA's Form 301.

OSHA's Form 300A (Rev. 01/2004)

Summary of Work-Related Injuries and Illnesses

Year 20__ __

U.S. Department of Labor
Occupational Safety and Health Administration

Form approved OMB no. 1218-0176

All establishments covered by Part 1904 must complete this Summary page, even if no work-related injuries or illnesses occurred during the year. Remember to review the Log to verify that the entries are complete and accurate before completing this summary.

Using the Log, count the individual entries you made for each category. Then write the totals below, making sure you've added the entries from every page of the Log. If you had no cases, write "0."

Employees, former employees, and their representatives have the right to review the OSHA Form 300 in its entirety. They also have limited access to the OSHA Form 301 or its equivalent. See 29 CFR Part 1904.35, in OSHA's recordkeeping rule, for further details on the access provisions for these forms.

Number of Cases

Total number of deaths	Total number of cases with days away from work	Total number of cases with job transfer or restriction	Total number of other recordable cases
____ (G)	____ (H)	____ (I)	____ (J)

Number of Days

Total number of days away from work	Total number of days of job transfer or restriction
____ (K)	____ (L)

Injury and Illness Types

Total number of . . .
(M)

(1) Injuries	____	(4) Poisonings	____
		(5) Hearing loss	____
(2) Skin disorders	____	(6) All other illnesses	____
(3) Respiratory conditions	____		

Post this Summary page from February 1 to April 30 of the year following the year covered by the form.

Public reporting burden for this collection of information is estimated to average 50 minutes per response, including time to review the instructions, search and gather the data needed, and complete and review the collection of information. Persons are not required to respond to the collection of information unless it displays a currently valid OMB control number. If you have any comments about these estimates or any other aspects of this data collection, contact: US Department of Labor, OSHA Office of Statistical Analysis, Room N-3644, 200 Constitution Avenue, NW, Washington, DC 20210. Do not send the completed forms to this office.

Establishment information

Your establishment name ____________________

Street ____________________

City ____________ State ____ ZIP ____

Industry description (*e.g., Manufacture of motor truck trailers*) ____________________

Standard Industrial Classification (SIC), if known (*e.g., 3715*)

__ __ __ __

OR

North American Industrial Classification (NAICS), if known (e.g., 336212)

__ __ __ __ __ __

Employment information (*If you don't have these figures, see the Worksheet on the back of this page to estimate.*)

Annual average number of employees ________

Total hours worked by all employees last year ________

Sign here

Knowingly falsifying this document may result in a fine.

I certify that I have examined this document and that to the best of my knowledge the entries are true, accurate, and complete.

Company executive ____________ Title ____________

( ) - Phone ____________ / / Date ____________

Figure 10-3. OSHA's Form 300A.

Reporting Requirements

Reporting Fatalities and Catastrophes

All employers covered by the Occupational Safety and Health Act of 1970 (P.L. 91-596) must report to OSHA any workplace incident resulting in a fatality or the in-patient hospitalization of three or more employees within eight (8) hours of the occurrence.

Keeping Injury and Illness Records

Unless requested by the BLS or OSHA, an employer does not have to keep injury and illness records if there are 10 or fewer employees during the entire previous calendar year, or if the business is classified in a specific low-hazard retail, service, finance, insurance, or real estate industry.

Non-exempt Employers' Duties

Employers that are not exempt from OSHA's recordkeeping requirements must prepare and maintain records of work-related injuries and illnesses. *Title 29 of The Code of Federal Regulations (CFR) Part 1904-* "Recording and Reporting Occupational Injuries and Illnesses," specifically identifies which cases to record.

- The Log of Work-Related Injuries and Illnesses (Form 300) is used to list injuries and illnesses and track days away from work, restricted, or transferred.
- The Injury and Illness Report (Form 301) is used to record supplementary information about recordable cases. A workers' compensation or insurance form can be used if it contains the same information.
- The Summary (Form 300A) is used to show totals for the year in each category. The summary is posted from February 1 to April 30 of each year.

State Plan States

A workplace in a state operating under an OSHA-approved plan must follow the state plan recordkeeping regulations of their particular state. These states may have some more stringent or supplemental requirements for reporting fatalities and catastrophes. Industry exemptions may also differ in states that operate under an OSHA-approved plan.

THE NEW RECORDKEEPING RULE

As previously noted, OSHA's revised rule took effect January 1, 2002. The rule addresses the recording and reporting of occupational injuries and illnesses that affect approximately 1.4 million establishments. And, as previously discussed, a number of specific industries in the retail, service, finance, insurance, and real estate sectors that are classified as low hazard are exempt from most requirements. Also exempt is the small business employer with 10 or fewer employees.

The new rule improves employee involvement, calls for greater employee privacy protection, creates simpler forms, provides clearer regulatory requirements, and allows employers more flexibility to use computers to meet OSHA regulatory requirements. A summary of the key provisions are:

- Updates three recordkeeping forms:
  - OSHA Form 300—The Log of Work-Related Injuries and Illnesses (Figure 10-1)
  - OSHA Form 301—The Injury and Illness Incident Reports (Figure 10-2)
  - OSHA Form 300A—Summary of Work-Related Injuries and Illness (Figure 10-3)

- Provides a single set of recording criteria for both work-related injuries and work-related illnesses. (The former rule required employers to record all illnesses, regardless of severity.)

- Requires records to include a work-related injury or illness resulting in one of the following:
  - Death
  - Days away from work
  - Restricted work or transfer to another job
  - Medical treatment beyond first aid
  - Loss of consciousness, or
  - Diagnosis of a significant injury or illness by a physician or other licensed health care professional.

- Includes new definitions of medical treatment, first aid, and restricted work to simplify recording decisions.

- Requires a significant degree of aggravation before a preexisting injury or illness is considered work-related.

- Adds further exceptions to the definition of work-relatedness to limit recording of cases involving eating and drinking of food and beverages, common colds and flu, blood donations, exercise programs, mental illnesses, etc.

- Clarifies the recording of "light duty" or restricted work cases, requires employers to record cases when the injured or ill employee is restricted from "routine job functions," which are defined as work activities the employee regularly performs at least once a week.

- Requires employers to record all needlestick and sharps injures involving contamination by another person's blood or other potentially infectious materials.

- Includes separate provisions describing the recording criteria for cases involving the work-related transmission of tuberculosis.

- Eliminates the term "lost workdays" and requires recording of days away from work or days restricted or days transferred to another job. Calls for employers to count calendar days rather than workdays.

- Requires employers to establish a procedure for employees to report injuries and illnesses and tell their employees how to report. (Employers are prohibited from discriminating against employees who do report by Section 11(c) of the *Occupational Safety and Health Act of 1970*.)

- Employees and former employees will be guaranteed access to their individual OSHA 301 forms. Employee representatives will be provided access to the "information about the case" section of the OSHA 301 form in establishments where they represent employees.

- Protects employee privacy by:

- Prohibiting employers from entering an individual's name on Form 300 for certain types of injuries or illnesses (e.g., sexual assaults, HIV infections, mental illnesses);
- Allowing employers not to describe the nature of sensitive injures where the employee's identity would be known
- Giving employee representatives access only to the portion of Form 301 that contains no personal information; and
- Requiring employer to remove employees' names before providing the data to persons not provided access rights under the rule.

- Requires the annual summary to be posted for three months instead of one. Requires certification of the summary by a company executive.

- Excludes some public transportation and motor vehicle accidents from the reporting of fatalities and catastrophes.

- States that operate their own job safety and health programs adopted comparable recordkeeping rules that were also effective January 1, 2002. States must have the same requirements for which injuries and illnesses are recordable and how they are recorded. However, other provisions, such as industry exemptions may be different as long as they are as stringent as the federal requirements.

MAJOR CHANGES

The major changes from OSHA's old 1904 recordkeeping rule to the new rule that employers began in using in 2002 are summarized here.

Scope

- The list of service and retail industries that are partially exempt from the rule has been updated. Some establishments that were covered under the old rule will not be required to keep OSHA records under the new rule and some formerly exempted establishments will now have to keep records. (§1904.2)

- The new rule continues to provide a partial exemption for employers who had 10 or fewer workers at all times in the previous calendar year. (§1904.1)

Forms

- The new OSHA Form 300 (Log of Work-Related Injuries and Illnesses) has been simplified and can be printed on smaller legal-sized paper (see Figure 10-1).
- The new OSHA Form 301 (Injury and Illness Incident Report) includes more data about how the injury or illness occurred (see Figure 10-2).
- The new OSHA Form 300A (Summary of Work-Related Injuries and Illnesses) provides additional data to make it easier for employers to calculate incidence rates (see Figure 10-3).
- Maximum flexibility has been provided so employers can keep all the information on computers, at a central location, or on alternative forms, as long as the information is compatible and the data can be produced when needed. (§1904.29 and §1904.30)

Work Related

- A "significant" degree of aggravation is required before a preexisting injury or illness becomes work-related. (§1904.5(a))
- Additional exceptions have been added to the geographic presumption of work relationship; cases arising from eating and drinking of food and beverages, blood donations, exercise programs, etc. no longer need to be recorded. Common cold and flu cases also no longer need to be recorded. (§1904.5(b)(2))
- Criteria for deciding when mental illnesses are considered work-related have been added. (§1904.5(b)(2))
- Sections have been added clarifying work relationship when employees travel or work out of their home. (§1904.5(b)(6) and §1904.5(b)(7))

Recording Criteria

- Different criteria for recording work-related injuries and work-related illnesses are eliminated; one set of criteria is used for both. The former rule required employers to record all illnesses, regardless of severity. (§1904.4)

- Employers are required to record work-related injuries or illnesses if they result in one of the following: death; days away from work; restricted work or transfer to another job; medical treatment beyond first aid; loss of consciousness; or diagnosis of a significant injury/illness by a physician or other licensed health care professional. (§1904.7(a))

- New definitions are included for medical treatment and first aid. First aid is defined by treatments on a finite list. All treatment not on this list is medical treatment. (§1904.7(b)(5))

- The recording of "light duty" or restricted work cases is clarified. Employers are required to record cases as restricted work cases when the injured or ill employee only works partial days or is restricted from performing their "routine job functions" (defined as work activities the employee regularly performs at least once weekly). (§1904.7(b)(4))

- Employers are required to record all needlestick and sharps injuries involving contamination by another person's blood or other potentially infectious material. (§1904.8)

- Musculoskeletal disorders (MSDs) are treated like all other injuries or illnesses: they must be recorded if they result in days away, restricted work, transfer to another job, or medical treatment beyond first aid.

- Special recording criteria are included for cases involving the work-related transmission of tuberculosis or medical removal under OSHA standards. (§1904.9 and §1904.11)

Day Counts

- The term "lost workdays" is eliminated and the rule requires recording of days away, days of restricted work, or transfer to another job. Also, new rules for counting that rely on calendar days instead of workdays are included. (§1904.7(b)(3))

- Employers are no longer required to count days away or days of restriction beyond 180 days. (§1904.7(b)(3))

- The day on which the injury or illness occurs is not counted as a day away from work or a day of restricted work. (§1904.7(b)(3) and §1904.7(b)(4))

Annual Summary

- Employers must review the 300 Log information before it is summarized on the 300A form. (§1904.32(a))

- The new rule includes hours worked data to make it easier for employers to calculate incidence rates. (§1904.32(b)(2))

- A company executive is required to certify the accuracy of the summary. (§1904.32(b)(3))

- The annual summary must be posted for three months instead of one. (§1904.32(b)(6))

Employee Involvement

- Employers are required to establish a procedure for employees to report injuries and illnesses and to tell their employees how to report. (§1904.35(a))

- The new rule informs employers that the OSH Act prohibits employers from discriminating against employees who do report. (§1904.36)

- Employees are allowed to access the 301 forms to review records of their own injuries and illnesses. (§1904.35(b)(2))

- Employee representatives are allowed to access those parts of the OSHA 301 form relevant to workplace safety and health. (§1904.35(b)(2))

Protecting Privacy

- Employers are required to protect employee's privacy by withholding an individual's name on Form 300 for certain types of sensitive injuries/illnesses (e.g., sexual assaults, HIV infections, mental illnesses, etc.). (§1904.29(b)(6) to §1904.29(b)(8))

- Employers are allowed to withhold descriptive information about sensitive injuries in cases where not doing so would disclose the employee's identity. (§1904.29(b)(9))

- Employee representatives are given access only to the portion of Form 301 that contains information about the injury or illness, while personal information about the employee and his or her health care provider is withheld. (§1904.35(b)(2))

- Employers are required to remove employees' names before providing injury and illness data to persons who do not have access rights under the rule. (§1904.29(b)(10))

Reporting Information to the Government

- Employers must call in all fatal heart attacks occurring in the work environment. (§1904.39(b)(5))
- Employers do not need to call in public street motor vehicle accidents except those in a construction work zone. (§1904.39(b)(3))
- Employers do not need to call in commercial airplane, train, subway or bus accidents. (§1904.39(b)(4))
- Employers must provide records to an OSHA compliance officer who requests them within 4 hours. (§1904.40(a))

SIDE-BY-SIDE COMPARISON OF THE RECORDKEEPING RULE

Some of the specific changes in the new rule include:

1. Changes in coverage;
2. The OSHA Forms;
3. The Recording Criteria in determination of work-relationship,
    a. elimination of different recording criteria for injuries and illnesses,
    b. days away and job restriction/ transfer,
    c. definition of medical treatment and first aid,
    d. recording of needlestick and sharps injuries, and
    e. recording of tuberculosis;
4. Change in ownership;
5. Employee involvement;
6. Privacy protections; and
7. Computerized and centralized records.

This listing is not comprehensive; it is intended only to give a general overview of the differences between the old rule and the new one. See Table 10-2 for a sibe-by-side comparison.

Table 10-2. Side by Side Comparison.

Old Rule	New Rule
Forms §1904.29	
OSHA 200 - Log and Summary OSHA 101 - Supplemental Record	OSHA 300 - Log OSHA 300A - Summary OSHA 301 - Incident Report
Work-Related §1904.5	
Any aggravation of a pre-existing condition by a workplace event or exposure makes the case work-related	***Significant*** aggravation of a pre-existing condition by a workplace event or exposure makes the case work-related
Exceptions to presumption of work relationship: 1) Member of the general public 2) Symptoms arising on premises totally due to outside factors 3) Parking lot/Recreational facility	Exceptions to presumption of work relationship: 1) Member of the general public 2) Symptoms arising on premises totally due to outside factors 3) Voluntary participation in wellness program 4) Eating, drinking and preparing one's own food 5) Personal tasks outside working hours 6) Personal grooming, self-medication, self infliction 7) Motor vehicle accident in parking lot/ access road during commute 8) Cold or flu 9) Mental illness unless employee voluntarily presents a medical opinion stating that the employee has a mental illness that is work- related.
New Case §1904.6	
New event or exposure, new case	Aggravation of a case where signs or symptoms have not resolved is a continuation of the original case
30 day rule for CTDs	No such criteria
General Recording Criteria §1904.7	
All work-related illnesses are recordable	Work-related illnesses are recordable if they meet the general recording criteria
Restricted work activity occurs if the employee: 1 Cannot work a full shift 2 Cannot perform all of his/her normal job duties, defined as any duty he/she would be expected to do throughout the calendar year.	Restricted work activity occurs if the employee: 1 Cannot work a full shift 2 Cannot perform all of his or her routine job functions, defined as any duty he or she regularly performs at least once a week
Restricted work activity limited to the day of injury makes case recordable	Restricted work activity limited to the day of injury does not make case recordable

(*Continued*)

Table 10-2. Side by Side Comparison (*Continued*).

Day counts: Count workdays No cap on count	Day Counts: Count Calendar days 180 day cap on count:
Medical treatment does not include: 1) Visits to MD for observation only 2) Diagnostic procedures 3) First aid	Medical treatment does not include: 1) Visits to MD for observation and counseling only 2) Diagnostic procedures (including administration of prescription medication for diagnostic purposes) 3) First aid
First Aid list in Bluebook was a list of examples and not comprehensive	First Aid list is comprehensive. Any other procedure is medical treatment.
2 doses prescription med - Medical Treatment (MT) Any dosage of OTC med - First Aid (FA) 2 or more hot/cold treatments - MT Drilling a nail - MT Butterfly bandage/Steri-Strip - MT	1 dose prescription med - MT OTC med at prescription strength - MT Any number of hot/cold treatments - FA Drilling a nail - FA Butterfly bandage/Steri-Strip – FA
Non-minor injuries recordable: 1) fractures 2) 2nd and 3rd degree burns	Significant diagnosed injury or illness recordable: 1) fracture 2) punctured ear drum 3) cancer 4) chronic irreversible disease
Specific disorders	
Hearing loss - Federal enforcement for 25dB shift in hearing from original baseline	Beginning 1/1/03 record all work-related hearing loss cases that meet BOTH of the following conditions on the same audiometric test for either ear: 1. The employee has experienced a Standard Threshold Shift (STS) **AND** 2. The employee's total hearing level is 25 dB or more above audiometric zero (averaged at 2000, 3000, & 4000 Hz) in the same ear(s) as the STS. Beginning 1/1/04 A separate hearing loss column on the OSHA 300 Log beginning in Calendar year 2004.
Needlesticks & 'sharps injuries' - Record only if case results in med treatment, days away, days restricted or sero-conversion	Needlesticks & 'sharps injuries' - Record all needlesticks & injuries that result from sharps potentially contaminated with another persons blood or other potentially infectious material
Medical removal under provisions of other OSHA standards - all medical removal cases recordable	Medical removal under provisions of other OSHA standards - all medical removal cases recordable

(*Continued*)

Table 10-2. Side by Side Comparison (*Concluded*).

TB - Positive skin test recordable when known workplace exposure to active TB disease. Presumption of work relationship in 5 industries	TB - Positive skin test recordable when known workplace exposure to active TB disease. No presumption of work relationship in any industry
Other issues	
Must enter the employees name on all cases	Must enter 'Privacy Cases' rather than the employee's name, & keep a separate list of the case number & corresponding names
Access - employee access to entire log, including names; No access to supplementary form (OSHA 101)	Access - employee and authorized representative access to entire log, including names; Employee access to individual's Incident Report (OSHA 301); Authorized Representative access to portion of all OSHA 301s
Fatality reporting - Report all work-related fatalities to OSHA	Fatality reporting - do not need to report fatalities resulting from motor vehicle accident on public street or highway that do not occur in construction zone
Certification - the employer, or the employee who supervised the preparation of the Log and Summary, can certify the annual summary	Certification - company executive must certify annual summary
Posting - post annual summary during month of February	Posting - Post annual summary from Feb 1 to April 30
No such requirement	Each employee must be informed on the injury/illness reporting procedure.

Source: OSHA

Occupational Safety and Health Definitions

Work-related illness and injuries are events or exposures in the work environment that caused or contributed to the condition or significantly aggravated a preexisting condition.

Recordable cases include work-related injuries and illnesses that result in:
Death
Loss of consciousness
Days away from work
Restricted work activity or job transfer
Medical treatment (beyond first aid)

Significant work related injuries or illnesses that are diagnosed by a physician or other licensed health care professional. These include any work related case involving cancer, chronic irreversible disease, a fractured or cracked bone, or a punctured eardrum.

(*Continued*)

Occupational Safety and Health Definitions (*Continued*)

Additional criteria that can result in a recordable case include:
Any needle-stick injury or cut from a sharp object that is contaminated with another person's blood or other potentially infectious material.

Any case requiring an employee to be medically removed under the requirements of an OSHA health standard.

Tuberculosis infection as evidenced by a positive skin test or diagnosis by a physician or other licensed health care professional after exposure to a known case of active tuberculosis.

An employee's hearing test (audiogram) reveals: 1) that the employee has experienced a Standard Threshold Shift (STS) in hearing in one or both ears (averaged at 2000, 3000, and 4000 Hz) and 2) the employee's total hearing level is 25 decibels (dB) or more above the audiometric zero (also averaged at 2000, 3000, and 4000 Hz) in the same ear(s) as the STS.

Days away from work, days of restricted work activity or job transfer (DART) are cases that involve days away from work, or days of restricted work activity, job transfer, or both.

Cases involving days away from work are cases requiring at least one day away from work with or without days of job transfer or restriction.

Job transfer or restriction cases occur when, as a result of a work-related injury/illness, an employer or health care professional keeps, or recommends keeping an employee from doing the routine functions of their job or from working the full workday that the employee would have been scheduled to work before the injury/illness occurred.

Other recordable cases are recordable cases that do not involve death, days away from work or days of restricted work activity or job transfer.

Incidence rate is the number of injuries and/or illnesses per 100 full-time workers and were calculated as: (N/EH) X 200,000 where:

N = number of injuries and/or illnesses

EH = total hours worked by all employees during the calendar year

200,000 = base for 100 full-time equivalent workers (working 40 hours per week, 50 weeks per year).

(*Continued*)

Occupational Safety and Health Definitions (*Concluded*)

Occupational injury is any wound or damage to the body resulting from an event in the work environment.

Source: U.S. Department of Labor, Bureau of Labor Statistics

Occupational illnesses:

Skin diseases or disorders are illnesses involving the worker's skin that are caused by work exposure to chemicals, plants or other substances. Examples: Contact dermatitis, eczema, or rash caused by primary irritants and sensitizers or poisonous plants; oil acne; friction blisters, chrome ulcers; inflammation of the skin.

Respiratory conditions are illnesses associated with breathing hazardous biological agents, chemicals, dust, gases, vapors, or fumes at work. Examples: Silicosis, asbestosis, pneumonitis, pharyngitis, rhinitis or acute congestion; farmer's lung, beryllium disease, tuberculosis, occupational asthma, reactive airways dysfunction syndrome (RADS), chronic obstructive pulmonary disease (COPD), hypersensitivity pneumonitis, toxic inhalation injury, such as metal fume fever, chronic obstructive bronchitis and other pneumoconioses.

Poisoning includes disorders evidenced by abnormal concentrations of toxic substances in blood, other tissues, other bodily fluids, or the breath that are caused by the ingestion or absorption of toxic substances into the body. Examples: Poisoning by lead, mercury, cadmium, arsenic, or other metals; poisoning by carbon monoxide, hydrogen sulfide, or other gases; poisoning by benzene, benzol, carbon tetrachloride, or other organic solvents; poisoning by insecticide sprays such as parathion or lead arsenate; poisoning by other chemicals such as formaldehyde.

Hearing loss. Noise-induced hearing loss for recordkeeping purposes is a change in hearing threshold relative to the baseline audiogram of an average of 10 dB or more in either ear at 2000, 3000, and 4000 hertz and the employee's total hearing level is 25 decibels (dB) or more above the audiometric zero (also averaged at 2000, 3000, and 4000 hertz) in the same ear(s).

All other occupational illnesses. Examples: Heatstroke, sunstroke, heat exhaustion, heat stress and other effects of environmental heat; freezing,

(*Continued*)

frostbite, and other effects of exposure to low temperatures; decompression sickness; effects of ionizing radiation (isotopes, x-rays, radium); effects of non-ionizing radiation (welding flash, ultra-violet rays, lasers); anthrax; bloodborne pathogenic diseases such as AIDS, HIV, hepatitis B or hepatitis C; brucellosis; malignant or benign tumors; histoplasmosis; coccidioidomycosis.

Source: U.S. Department of Labor, Bureau of Labor Statistics

Case Characteristics:

Nature of injury or illness names the principal physical characteristic of a disabling condition, such as sprain/strain, cut/laceration, or carpal tunnel syndrome.

Part of body affected is directly linked to the nature of injury or illness cited, for example, back sprain, finger cut, or wrist and carpal tunnel syndrome.

Source of injury or illness is the object, substance, exposure, or bodily motion that directly produced or inflicted the disabling condition cited. Examples are a heavy box, a toxic substance, fire/flame, and bodily motion of injured/ill worker.

Event or exposure—signifies the manner in which the injury or illness was produced or inflicted, for example, overexertion while lifting or fall from ladder.

Median days away from work—This is the measure used to summarize the varying lengths of absences from work among the cases with days away from work. Half the cases involved more days and half involved less days than a specified median.

Source: U.S. Department of Labor, Bureau of Labor Statistics

An injury or illness is considered work-related if an event or exposure in the work environment caused or contributed to the condition or significantly aggravated a preexisting condition. Work-relatedness is presumed for injuries and illnesses resulting from events or exposures occurring in the workplace, unless an exception specifically applies. See 29

CFR Part 1904.5(b)(2) for the exceptions. The work environment includes the establishment and other locations where one or more employees are working or are present as a condition of their employment.

Any significant work-related injury or illness that is diagnosed by a physician or other licensed health care professional must be recorded. Any work-related case involving cancer, chronic irreversible disease, a fractured or cracked bone, or a punctured eardrum must also be recorded.

The following conditions must be recorded when they are work-related:

- any needlestick injury or cut from a sharp object that is contaminated with another person's blood or other potentially infectious material;

- any case requiring an employee to be medically removed under the requirements of an OSHA health standard; tuberculosis infection as evidenced by a positive skin test or diagnosis by a physician or other licensed health care professional after exposure to a known case of active tuberculosis.

- an employee's hearing test (audiogram) reveals 1) that the employee has experienced a Standard Threshold Shift (STS) in hearing in one or both ears (averaged at 2000, 3000, and 4000 Hz) and 2) the employee's total hearing level is 25 decibels (dB) or more above audiometric zero (also averaged at 2000, 3000, and 4000 Hz) in the same ear(s) as the STS.

Medical treatment includes managing and caring for a patient for the purpose of combating disease or disorder. The following are not considered medical treatments and are NOT recordable:

- visits to a doctor or health care professional solely for observation or counseling;

- diagnostic procedures, including administering prescription medications that are used solely for diagnostic purposes; and

- any procedures that can be labeled first aid.

First Aid

If the incident required only the following types of treatment, it is considered first aid. These cares are not to be recorded if it involves only:

- Using non-prescription medications at non-prescription strength;
- Administering tetanus immunizations;
- Cleaning, flushing, or soaking wounds on the skin surface;
- Using wound coverings, such as bandages, Band Aids™;
- Using hot or cold therapy; using any totally non-rigid means of support, such as elastic bandages, wraps, non-rigid back belts, etc;
- Using temporary immobilizations devices while transporting an accident victim—splints, slings, neck collars, or back boards;
- Drilling a fingernail or toenail to relieve pressure, or draining fluids from a blister;
- Using eye patches' using simple irrigation or a cotton swab to remove foreign bodies not embedded in or adhered to the eye;
- Using irrigation, tweezers, cotton swab or other simple means to remove splinters or foreign material from areas other than the eye;
- Using finger guards;
- Using massages; and
- Drinking fluids to relieve heat stress.

Sources

U.S. Department of Labor Occupational Safety and Health Administration, *Recordkeeping—It's New, It's Improved, and It's Easier,* OSHA 3169, 2001.

U.S. Department of Labor, Occupational Safety and Health Administration, *OSHA Forms for Recording Work-Related Injuries and Illnesses,* New OSHA 300 Form 1-01-04.

U.S. Department of Labor, Occupational Safety and Health Administration, *OSHA Recordkeeping Handbook,* OSHA 3245-09R, 2005.

U.S. Department of Labor, Occupational Safety and Health Administration, *Recordkeeping, It's new, it's improved, and it's easier…,* OSHA 3169, 2001.

U.S. Department of Labor, Occupational Safety and Health Administration, *Recordkeeping,* OSHA Fact Sheet, 2001.

Chapter 11

Change Management

Change management is a loosely defined term that refers to a broad array of activities and initiatives that occur in the workplace. As such, in order to be effective, a change management program must integrate those program elements that address any of the variety of elements: communication, training and testing, program planning, market analysis and implementation of new policies and procedures.

Adjusting to the changes in the workplace is the focus of a change management program. A good change management strategy includes:

- Communication
- Discussion
- Involvement

Communication

In any business setting, a good change management program must include communication. Communication is at the core of the change management program. The change itself should be communicated to the tenant/occupant/employee along with the hows, whys and wherefores. Prerequisites to an effective change management approach are:

- Definition
- Explanation
- Progression

Definition

Defining what the change is an essential aspect of the change management strategy. For example, define the major goal of the project. Is it to support current business needs, improve customer service, or attract and retain employees--the employee/tenant/occupant should know the reason for the change.

Explanation

Explaining how it will occur and why it needs to occur, as well as describing when it well occur is another important part of the communication process.

Progression

Providing the timely information on the various stages of implementation are prerequisites to an effective change management approach.

Discussion

Employees/occupants/tenants may be anxious, fearful, etc. about the change that is about to occur. Discussing what their role and integration into the change should alleviate the concerns of the employee/occupant/tenant. With employees, for example, discussing such issues as career development, performance assessment, technical skills and work and family issues should alleviate some of their concerns. For tenant/occupants, issues such as how the change will impact upon their operations are an issue that must be discussed. In fact, bringing the tenants/occupants into the change process is one of the most effective ways to secure their commitment and support. A planning committee comprised of the building owners/executives, company manager and "rank-and-file" serves an additional purpose. New suggestions and new ideas for implementing change can be generated from such a group.

Involvement

As discussed previously, involving employees/tenants/occupants in the planning and design; communicating the change, information about the change including updating the employee/tenant/occupant about the change; and, finally, soliciting their input, should provide greater support and commitment on their part to the change.

DEVELOPING A COMMUNICATIONS STRATEGY

A second aspect to the change management strategy includes developing a communications strategy.

Facility managers can use a number of methods to plan and implement a communications strategy for all aspects of their change management initiative. Rooted in the principles of marketing, these methods involve:

- Market analysis.
- Program planning.
- Promotion.

Each of these factors, in turn, involves:

- Accurate awareness of a company's change management needs.
- Adapting to an organizational mandate.
- Organized communications systems.
- Cost and quality control.

Market Analysis

Effective "market analysis" begins by focusing on the company's current operating environment to identify safety and health hazards and potential hazards that are present in the workplace. Market analysis provides the necessary information for addressing hazards, vis-à-vis program planning. Good market research allows companies to make the most efficient use of their resources and measures success in safety and health communications

Market analysis identifies the communication need.

Program Planning

Program planning is product development—the process of bringing staff, resources (internal and external) and capacity together to meet the company's safety and health needs, as well as ensure compliance with regulatory mandates. Like facility planning, effective program planning (or product development) expedites bringing new programs, products and services onboard, while assuring the cost-efficiency of the result.

Program planning develops the structure and systems to deliver communications and satisfy the facility's change management requirements.

Promotion

The promotional aspects of marketing serve as a logical follow-up to program planning. These promotional aspects facilitate a practical understanding of the company's change management program.

Promotion is the active process of communicating safety policies, training and other important safety and health information to the building occupants.

EFFECTIVENESS OF THE COMMUNICATIONS STRATEGY

The effectiveness of the communications strategy is determined by how well it facilitates a practical understanding of the change management initiative, as well as the responsibilities that each employee assumes.

The specific criteria used to evaluate effectiveness include the right combination of structure, systems and staff skills.

Structure

By nature, market and program planning cross all organizational lines and require the input and commitment of all employee levels and analysis become the functions of all employees, with each person assuming responsibilities for change management and acting upon those responsibilities.

Effective program planning, like facilities planning, requires a skilled integrator—the facility manager, who is in charge and responsible for the results of the company's change management initiative. Facility managers must work with the Change Planning Committee, the organizationally sanctioned group that has the authority to make and enforce decisions.

Systems

All of the company's operating systems must be utilized to implement marketing and program planning strategies, including market analysis, budgeting, financial planning, legal resources and human resources.

Managers and supervisors are responsible for ensuring the successful change in their respective divisions and departments. Facility managers have the prime responsibility for the company's overall change management initiative. All managers and supervisors should incorporate practical considerations in their specific communications strategies.

Staff Skills

Staff skills, including analysis, consensus-building and process management must be utilized to fully implement the company's change management communication strategy. The practical considerations for doing so are described below.

MAKING THE STRATEGY WORK

Making any change management program work effectively involves a number of communications and management techniques. They include:

- Developing "open" lines of communication.
- Communication incentive.
- Management techniques.

Developing Open Lines of Communication

Open communications lines provide for a "free" flow of information, ideas and problem-solving techniques. Developing these open lines is essential to the change management strategy. Examples include meetings, change management committees, company newsletters and "change" hotlines.

Meetings

Multilevel meetings provide the vehicles for informing employees/tenants/occupants at all levels of the change issues and the current developments. Regularly scheduled peer-level meetings, as well as manager-employee meetings, give everyone in the company an opportunity to exchange ideas, concerns, and the opportunity to provide input into company direction and policy.

"Change" Committees

A formal change management committee, comprised of management and "rank and file" representatives should be instituted, and meet on a regularly scheduled basis to discuss problems and concerns.

Committee actions, recommendations, etc., should be "publicized" so that all employees are updated on a company's safety and health efforts.

Company Newsletters

Every newsletter should contain a "change section" that, again, keeps employees apprised of the organization's change management efforts.

Recognizing divisions, departments and individual employees for their own particular "change initiatives/implementations" keeps the change issue in the forefront, while giving everyone an opportunity to "share the spotlight.

"Change" Hotlines

To ensure effective communication bases with corporate offices, multi-site companies can consider establishing a hotline, so that any serious problems are addressed immediately.

Communication Incentives

Communication incentives are non-financial opportunities that can be used to let employees know their worth to the organization. Some of these incentives include "change" recognition awards programs and/or luncheons where employees are publicly acknowledged for their safety efforts. Employee profiles in annual reports, shareholder notices, or client mailings are other techniques that employers may want to consider in recognizing the change-committed employee.

Sources

Gustin, Joseph F., *Safety Management, A Guide for Facility Managers*, UpWord Publishing, New York, 1996.

U.S. General Services Administration, Office of Governmentwide Policy, Office of Real Property, *The Integrated Workplace, A Comprehensive Approach to Developing Workspace*, May, 1991. Second Printing, April, 2000.

Appendix I
Directory of Sources

A list of publications, trade associations and approved sources of OSHA-funded consultation along with current ADA information is provided below. Websites are also included for more information.

Publications

Building Operating Management Magazine
www.facilitiesnet.com

Buildings Magazine
www.buildings.com

Business Facilities Magazine
www.businessfacilities.com

Facility Manager Magazine
www.todaysfacilitymanager.com

Health Facilities Management Magazine
800.242.4890
www.hfmagazine.com

Journal of Property Management Magazine
www.irem.org

Maintenance Solutions
www.facilitiesnet.com

Occupational Hazards Magazine
www.occupationalhazards.com

Occupational Health & Safety Magazine
www.oshonline.com

Safety and Health Magazine
1121 Spring Lake Drive
Itasca, IL 60143-3201
630.285.1121
630.285.1315 (FAX)
www.nsc.org

Security Magazine
2401 W. Big Beaver Rd., Suite 700
Troy, Michigan 48084
www.securitymagazine.com

Security Technology & Design Magazine
www.securityinfowatch.com

Today's Facility Manager Magazine
44 Apple Street Suite 3
Tinton Falls, NJ 07724
732.842.7433
732.758.6634 (FAX)
www.todaysfacilitymanager.com

Professional and Trade Organizations

The American Society of Safety Engineers
1800 E. Oakton St
Des Plaines, IL 60018
847.699.2929 (8:30 - 5:00 CST)
847.768.3434 (24 Hours) (FAX)
www.asee.org

Association of Health Facility Survey Agencies
5105 Solemn Grove Road
Garner, NC 27529
www.ahfsa.org

Association of Higher Education Facilities Officers
1643 Prince Street
Alexandria, VA 22314-2818
703.684.1446
www.appa.org

Council of Educational Facility Planners, International
The School Building Association
9180 E. Desert Cove Drive
Suite 104
Scottsdale, AZ 85260
480.391.0840
480.391.0940(FAX)
www.cefpi.org

Environmental Information Association
6935 Wisconsin Avenue, Suite 306
Chevy Chase, MD 20815-6112
301.961.4999
301.961.3094 (FAX)
www.eia-usa.org

Federal Facilities Council
500 Fifth St., N.W.
Washington, DC 20001
www..nationalacademies.org

International Facility Management Association
1 E. Greenway Plaza, Suite 1100
Houston, TX 77046-0194
713.623.4362
713.623.6124
www.ifma.org

National Association of State Facilities Administrators
NASFA
PO Box 11910
Lexington, KY 40578-1910
859.244.8181
859.244.8001
www.nasfa.net

National Constructors Association
AGC of America
2300 Wilson Boulevard
Suite 400
Arlington, VA 22201
703.548.3118 (phone)
703.548.3119 (FAX)
www.agc.org

National Fire Protection Association
1 Batterymarch Park,
Quincy, MA 02169-7471
617.770.3000
617.770.0700 (FAX)
www.nfpa.org

National Safety Council
1121 Spring Lake Drive
Itaska, IL 60143-3201
630.285.1121
630.285.1315 (FAX)
www.nsc.org

National Safety Management Society
P.O. Box 4460
Walnut Creek, CA 954596-0460
800.321.2910
www.nsms.us

OSHA Directory

REGIONAL OFFICES

In case of emergency call 1-800-321-OSHA

Region 1 *(Connecticut, Maine, Massachusetts, New Hampshire, Rhode Island, Vermont)*
Regional Office
JFK Federal Building, Room E340

Boston, Massachusetts 02203
617.565.9860
617.565.9827 FAX

Region 2 *(New Jersey, New York, Puerto Rico, Virgin Islands)*
Regional Office
201 Varick Street, Room 670
New York, New York 10014
212.337.2378
212.337.2371 FAX

Region 3 *(District of Columbia, Delaware, Maryland, Pennsylvania, Virginia, West Virginia)*
Regional Office
U.S. Department of Labor/OSHA
The Curtis Center-Suite 740 West
170 S. Independence Mall West
Philadelphia, PA 19106-3309
215.861.4900
215.861.4904 FAX

Region 4 *(Alabama, Florida, Georgia, Kentucky, Mississippi, North Carolina, South Carolina, Tennessee)*
Regional Office
61 Forsyth Street, SW
Atlanta, Georgia 30303
404.562.2300
404.562.2295 FAX

Region 5 *(Illinois, Indiana, Michigan, Minnesota, Ohio, Wisconsin)*
Regional Office
230 South Dearborn Street, Room 3244
Chicago, Illinois 60604
312.353.2220
312.353.7774 FAX

Region 6 *(Arkansas, Louisiana, New Mexico, Oklahoma, Texas)*
Regional Office
525 Griffin Street, Room 602

Dallas, Texas 75202
972.850.4145
972.850.4149 FAX

Region 7 *(Iowa, Kansas, Missouri, Nebraska)*
Regional Office
City Center Square
1100 Main Street, Suite 800
Kansas City, Missouri 64105
816.426.5861
816.426.2750 FAX

Region 8 *(Colorado, Montana, North Dakota, South Dakota, Utah, Wyoming)*
Regional Office
1999 Broadway, Suite 1690
Denver, Colorado 80202
720.264.6550
720.264.6585 FAX

Region 9 *(Arizona, California, Guam, Hawaii, Nevada) For issues involving federal agencies or private companies working for federal agencies in* ***Arizona, California, Guam, Hawaii****, and* ***Nevada****, call the numbers listed below. For issues involving* ***private*** *or* ***state government*** *employers in these states, refer to the appropriate* ***state office*** *in* ***Arizona, California, Hawaii****, and* ***Nevada.***

Region IX Federal Contact Numbers
71 Stevenson Street, Room 420
San Francisco, California 94105
415.975.4310
415.975.4319 FAX

Region 10 *(Alaska, Idaho, Oregon, Washington)*
Regional Office
1111 Third Avenue, Suite 715
Seattle, Washington 98101-3212
206.553.5930
206.553.6499 FAX

LOCAL AREA OFFICES

ALABAMA
Birmingham Area Office
950 22nd Street North
Suite 1050
Birmingham, AL 35203
205.731.1534
205.731.0504 FAX

Mobile Area Office
1141 Montlimar Drive, Suite 1006
Mobile Alabama 36609
251.441.6131
251.441.6396 FAX

ALASKA
Anchorage Area Office
U.S. Department of Labor - OSHA
222 W. 7th Avenue, Box 22
Anchorage, AK 99513
907.271.5152
907.271.4238 FAX

ARIZONA
Phoenix Office
800 W. Washington Street
Phoenix, AZ 85007
602.542.4654

Tucson Office
2675 East Broadway
Tucson, AZ 85716
520.628.5188

ARKANSAS
Little Rock Area Office
10810 Executive Center Dr
Danville Bldg #2; Ste 206

Little Rock, AR 72211
501.224.1841
501.224.4431 FAX

CALIFORNIA
California Department of Industrial Relations
1515 Clay Street, Suite 1901
Oakland, California 94612
415.703.5050
415.703.5058 FAX

COLORADO
Denver Area Office
1391 Speer Boulevard, Suite 210
Denver, Colorado 80204-2552
303.844.5285
303.844.6676 FAX

Englewood Area Office
7935 East Prentice Avenue, Suite 209
Englewood, Colorado 80111-2714
303.843.4500
303.843.515 FAX

CONNECTICUT
Bridgeport Area Office
Clark Building
1057 Broad Street, 4th Floor
Bridgeport, Connecticut 06604
203.579.5581
203.579.5516 FAX

Hartford Area Office
Federal Building
450 Main Street, Room 613
Hartford, Connecticut 06103
860.240.3152
860.240.3155 FAX

DELAWARE

Wilmington Area Office
Mellon Bank Building, Suite 900
919 Market Street
Wilmington, Delaware 19801
302.573.6518
302.573.6532 FAX

FLORIDA

Fort Lauderdale Area Office
8040 Peters Road, Building H-100
Fort Lauderdale, Florida 33324
954.424.0242
954.424.3073 FAX

Jacksonville Area Office
Ribault Building, Suite 227
1851 Executive Center Drive
Jacksonville, Florida 32207
904.232.2895
904.232.1294 FAX

Tampa Area Office
5807 Breckenridge Parkway, Suite A
Tampa, Florida 33610-4249
813.626.1177
813.626.7015 FAX

GEORGIA

Atlanta East Area Office
LaVista Perimeter Office Park
2183 N. Lake Parkway, Building 7
Suite 110
Tucker, Georgia 30084-4154
770.493.6644
770.493.7725 FAX

Atlanta West Area Office
2400 Herodian Way, Suite 250

Smyrna, Georgia 30080-2968
770.984.8700
770.984.8855 FAX

Savannah Area Office
450 Mall Boulevard, Suite J
Savannah, Georgia 31406
912.652.4393
912.652.4329 FAX

HAWAII
Hawaii Department of Labor and Industrial Relations
830 Punchbowl Street
Honolulu, Hawaii 96813
808.586.8844
808.586.9099 FAX

IDAHO
Boise Area Office
1150 North Curtis Road, Suite 201
Boise, Idaho 83706
208.321.2960
208.321.2966 FAX

ILLINOIS
Calumet City Area Office
1600 167th Street, Suite 9
Calumet City, Illinois 60409
708.891.3800
708.862.9659 FAX

Chicago North Area Office
701 Lee Street - Suite 950
Des Plaines, Illinois 60016
847.803.4800
847.390.8220 FAX

Fairview Heights District Office
11 Executive Drive, Suite 11

Fairview Heights, Illinois 62208
618.632.8612
618.632.5712 FAX

North Aurora Area Office
365 Smoke Tree Plaza
North Aurora, IL 60542
630.896.8700
630.892.2160 FAX

Peoria Area Office
2918 W. Willows Knolls Road
Peoria, Illinois 61614
309.589.7033
309.589.7326 FAX

INDIANA
Indianapolis Area Office
46 East Ohio Street, Room 423
Indianapolis, Indiana 46204
317.226.7290
317.226.7292 FAX

IOWA
U S Department of Labor
Occupational Safety and Health Administration
210 Walnut St Room 815
Des Moines IA 50309-2015
515.284.4794
515.284.4058 FAX

KANSAS
Wichita Area Office
271 W. 3rd Street North, Room 400
Wichita, KS 67202
316.269.6644
316.269.6185 FAX
Toll Free (Kansas Residents Only): 1.800.362.2896

KENTUCKY

Frankfort Area Office
John C. Watts Federal Office Building
330 West Broadway, Room 108
Frankfort, Kentucky 40601-1922
502.227.7024
502.227.2348 FAX

LOUISIANA

Baton Rouge Area Office
9100 Bluebonnet Centre Blvd, Suite 201
Baton Rouge, Louisiana 70809
225.298.5458
225.298.5457 FAX

MAINE

Bangor District Office
202 Harlow Street, Room 240
Bangor, ME 04401
207.941.8177
207.941.8179 FAX

Augusta Area Office
E.S. Muskie Federal Bldg
40 Western Ave., Room G-26
Augusta, ME 04330
207.626.9160
207.622.8213 FAX

MARYLAND

Baltimore/Washington Area Office
1099 Winterson Road
Suite 140
Linthicum, Maryland 21090
410.865.2055/2056
410.865.2068 FAX

MASSACHUSETTS

North Boston Area Office
Valley Office Park

13 Branch Street
Methuen, Massachusetts 01844
617.565.8110
617.565.8115 FAX

South Boston Area Office
639 Granite Street, 4th Floor
Braintree, Massachusetts 02184
617.565.6924
617.565.6923 FAX

Springfield Area Office
1441 Main Street, Room 550
Springfield, Massachusetts 01103-1493
413.785.0123
413.785.0136 FAX

MICHIGAN
Lansing Area Office
U.S. Department of Labor
Occupational Safety and Health Administration
315 West Allegan
Room 207
Lansing, Michigan 48933
517.487.4996
517.487.4997 FAX

MINNESOTA
The federal OSHA Area Office in Minneapolis was closed effective March 31, 2005. For federal and state-plan monitoring activity in Minnesota, contact:
Eau Claire Area Office
1310 W. Clairemont Avenue
Eau Claire, Wisconsin 54701
715.832.9019
715.832.1147 FAX

MISSISSIPPI
Jackson Area Office
3780 I-55 North, Suite 210

Jackson, Mississippi 39211-6323
601.965.4606
601.965.4610 FAX

MISSOURI
Kansas City Area Office
6200 Connecticut Avenue, Suite 100
Kansas City, Missouri 64120
816.483.9531
816.483.9724 FAX
Toll Free (Missouri Residents Only): 1.800.892.2674

St. Louis Area Office
911 Washington Avenue, Room 420
St. Louis, Missouri 63101
314.425.4249
314.425.4289 FAX
Toll Free (Missouri Residents Only): 1.800.392.7743

MONTANA
Billings Area Office
2900 4th Avenue North, Suite 303
Billings, Montana 59101
406.247.7494
406.247.7499 FAX

NEBRASKA
Omaha Area Office
Overland-Wolf Building
6910 Pacific Street, Room 100
Omaha, Nebraska 68106
402.553.0171
402.551.1288 FAX
Toll Free (Nebraska Residents Only): 1.800.642.8963

NEVADA
Nevada Division of Industrial Relations
400 West King Street
Carson City, Nevada 89073

775.684.7260
775.687.6305 FAX

NEW HAMPSHIRE
Concord Area Office
J.C. Cleveland Federal Bldg
53 Pleasant Street, Room 3901
Concord, New Hampshire 03301
603.225.1629
603.225.1580 FAX

NEW JERSEY
Avenel Area Office
1030 St. Georges Avenue
Plaza 35, Suite 205
Avenel, New Jersey 07001
732.750.3270
732.750.4737 FAX

Hasbrouck Heights Area Office
500 Route 17 South
2nd Floor
Hasbrouck Heights, New Jersey 07604
201.288.1700
201.288.7315 FAX

Marlton Area Office
Marlton Executive Park, Building 2
701 Route 73 South, Suite 120
Marlton, New Jersey 08053
856.396.2594
856.396.2593 FAX

Parsippany Area Office
299 Cherry Hill Road, Suite 103
Parsippany, New Jersey 07054
973.263.1003
973.299.7161 FAX

NEW MEXICO
The Albuquerque, NM, OSHA Area Office has been closed. Contact:
Lubbock Area Office
Federal Office Building
1205 Texas Avenue, Room 806
Lubbock, Texas 79401
806.472.7681/7685
806.472.7686 FAX

NEW YORK
Albany Area Office
401 New Karner Road, Suite 300
Albany, New York 12205-3809
518.464.4338
518.464.4337 FAX

Queens District Office of the Manhattan Area Office
45-17 Marathon Parkway
Little Neck, NY 11362
718.279.9060
718.279.9057 FAX

Buffalo Area Office
U. S. Dept. of Labor/OSHA
130 S. Elmwood Avenue, Suite 500
Buffalo, NY 14202-2465
716.551.3053
716.551.3126 FAX

Long Island Area Office
1400 Old Country Road
Suite 208
Westbury, New York 11590
516.334.3344
516.334.3326 FAX

Manhattan Area Office
201 Varick Street RM. 908
New York, NY 10014
212.620.3200
212.620.4121 (FAX)

Syracuse Area Office
3300 Vickery Road
North Syracuse, New York 13212
315.451.0808
315.451.1351 FAX

Tarrytown Area Office
660 White Plains Road, 4th Floor
Tarrytown, New York 10591-5107
914.524.7510
914.524.7515 FAX

NORTH CAROLINA
Raleigh Area Office
4407 Bland Road, Suite 210
Raleigh, North Carolina 27609
919.790.8096
919.790.8224 FAX

NORTH DAKOTA
Bismarck Area Office
Federal Office Building
1640 East Capitol Avenue
Bismarck, North Dakota 58501
701.250.4521
701.250.4520 FAX

OHIO
Cincinnati Area Office
36 Triangle Park Drive
Cincinnati, Ohio 45246
513.841.4132
513.841.4114 FAX

Cleveland Area Office
Federal Office Building
1240 East 9th Street, Room 899
Cleveland, Ohio 44199
216.615.4266
216.615.4234 FAX

Columbus Area Office
Federal Office Building
200 North High Street, Room 620
Columbus, Ohio 43215
614.469.5582
614.469.6791 FAX

Toledo Area Office
Ohio Building
420 Madison Avenue, Suite 600
Toledo, Ohio 43604
419.259.7542
419.259.6355 FAX

OKLAHOMA
Oklahoma City Area Office
55 North Robinson - Suite 315
Oklahoma City, Oklahoma 73102-9237
405.278.9560
405.278.9572 FAX

OREGON
Portland Area Office
Federal Office Building
1220 Southwest 3rd Avenue, Room 640
Portland, Oregon 97204
503.326.2251
503.326.3574 FAX

PENNSYLVANIA
Allentown Area Office
850 North 5th Street
Allentown, Pennsylvania 18102-1731
610.776.0592
610.776.1913 FAX

Erie Area Office
1128 State Street
Suite 200

Erie, PA 16501
814.461.1492
814.461.1498 FAX

Harrisburg Area Office
Progress Plaza
49 North Progress Avenue
Harrisburg, Pennsylvania 17109-3596
717.782.3902
717.782.3746 FAX

Philadelphia Area Office
US Custom House, Room 242
Second & Chestnut Street
Philadelphia, Pennsylvania 19106-2902
215.597.4955
215.597.1956 FAX

Pittsburgh Area Office
Federal Office Building, Room 1428
1000 Liberty Avenue
Pittsburgh, Pennsylvania 15222-4101
412.395.4903
412.395.6380 FAX

Wilkes-Barre Area Office
The Stegmaier Building
Suite 410
7 North Wilkes-Barre Boulevard
Wilkes-Barre, PA 18702-5241
570.826.6538
570.821.4170 FAX

PUERTO RICO
Puerto Rico Area Office
Triple S Building
1510 FD Roosevelt Avenue
Guaynabo, Puerto Rico 00968
787.277.1560
787.277.1567 FAX

RHODE ISLAND
Providence Area Office
Federal Office Building
380 Westminster Mall, Room 543
Providence, Rhode Island 02903
401.528.4669
401.528.4663 FAX

SOUTH CAROLINA
Columbia Area Office
1835 Assembly Street, Room 1472
Columbia, South Carolina 29201-2453
803.765.5904
803.765.5591 FAX

SOUTH DAKOTA
Regional Office
1999 Broadway, Suite 1690
Denver, Colorado 80202
720.264.6550
720.264.6585 FAX

TENNESSEE
Nashville Area Office
2002 Richard Jones Road, Suite C-205
Nashville, Tennessee 37215-2809
615.781.5423
615.781.5426 FAX

TEXAS
Austin Area Office
1033 La Posada Dr., Suite 375
Austin, Texas 78752-3832
512.374.0271
512.374.0086 FAX

Corpus Christi Area Office
Wilson Plaza, Suite 700
606 N Carancahua

Corpus Christi, Texas 78476
361.888.3420
361.888.3424 FAX

Dallas Area Office
8344 East RL Thornton Freeway
Suite 420
Dallas, Texas 75228
214.320.2400
214.320.2598 FAX

El Paso District Office
U.S. Dept. of Labor - OSHA
4849 N. Mesa St.
Suite 200
El Paso, TX 79912-5936
915.534.6251/6252
915.534.6259 FAX

Fort Worth Area Office
8713 Airport Freeway
Suite 302
Fort Worth, Texas 76180-7610
817.428.2470
817.581.7723 FAX

Houston North Area Office
507 North Sam Houston Parkway East
Suite 400
Houston, Texas 77060
281.591.2438
281.999.7457 FAX

Houston South Area Office
17625 El Camino Real, Suite 400
Houston, Texas 77058
281.286.0583/0584
281.286.6352 FAX

Lubbock Area Office
Federal Office Building
1205 Texas Avenue, Room 806
Lubbock, Texas 79401
806.472.7681/7685
806.472.7686 FAX

San Antonio District Office
800 Dolorosa St., Suite 203
San Antonio, TX 78207-4559
210.472.5040
210.472.5045 FAX

UTAH
Federal jurisdiction of Utah is administered by
U.S. Department of Labor
Occupational Safety and Health Administration
1391 Speer Blvd, Suite 210
Denver, CO 80204-2552
303.844.5285, Ext. 106

VERMONT
There are no Area Offices located in Vermont.
Contact the Regional Office for assistance.
JFK Federal Building, Room E340
Boston, Massachusetts 02203
617.565.9860
617.565.9827 FAX

VIRGINIA
Norfolk Area Office
Federal Office Building, Room 614
200 Granby Mall
Norfolk, Virginia 23510-1819
757.441.3820
757.441.3594 FAX

VIRGIN ISLANDS
There are no Federal OSHA Area Offices located in the Virgin Islands. Contact the Regional Office for assistance.

Regional Office
201 Varick Street, Room 670
New York, New York 10014
212.337.2378
212.337.2371 FAX

WASHINGTON
Bellevue Area Office
505 106th Avenue NE, Suite 302
Bellevue, Washington 98004
425.450.5480
425.450.5483 FAX

WASHINGTON DC
BALTIMORE/WASHINGTON AREA OFFICE
1099 Winterson Road
Suite 140
Linthicum, Maryland 21090
410.865.2055/2056
410.865.2068 FAX

WEST VIRGINIA
Charleston Area Office
405 Capitol Street
Suite 407
Charleston, West Virginia 25301-1727
304.347.5937
304.347.5275 FAX

WISCONSIN
Appleton Area Office
1648 Tri Park Way
Appleton, Wisconsin 54914
920.734.4521
920.734.2661 FAX

Eau Claire Area Office
1310 W. Clairemont Avenue
Eau Claire, Wisconsin 54701
715.832.9019
715.832.1147 FAX

Madison Area Office
4802 E. Broadway
Madison, Wisconsin 53716
608.441.5388
608.441.5400 FAX

Milwaukee Area Office
Henry S. Reuss Building, Suite 1180
310 West Wisconsin Avenue
Milwaukee, Wisconsin 53203
414.297.3315
414.297.4299 FAX

WYOMING
There are no Area Offices located in Wyoming. Contact the Region 8 Regional Office for assistance.
1999 Broadway, Suite 1690
Denver, Colorado 80202
720.264.6550
720.264.6585 FAX

OSHA Recordkeeping Coordinators

Region 1—(CT, MA, ME, NH, RI, VT)
Technical Assistance
U.S. DOL/OSHA
JFK Federal Bldg., Room E340
Boston, MA 02203
617.565.9856
617.565.9827 FAX

Region 2—(NJ, NY, PR, VI)
Technical Assistance
U.S. DOL/OSHA
201 Varick Street, Room 670
New York, NY 10014
212.337.2339
212.337.2371 FAX

Region 3—(DC, DE, MD, PA, VA, WV)
Technical Assistance
U.S. DOL/OSHA
Gateway Building - Suite 2100
3535 Market Street
Philadelphia, PA 19104
215.861.4900
215.861.4904 FAX

Region 4—(AL, FL, GA, KY, MS, NC, SC, TN)
Technical Assistance
U.S. DOL/OSHA
Sam Nunn Atlanta Federal Center
61 Forsyth Street, SW - Room 6T50
Atlanta, GA 30303
404.562.2279
404.562.2295 FAX

Region 5—(IL, IN, MI, MN, OH, WI)
Technical Assistance
U.S. DOL/OSHA
230 South Dearborn, Room 3244
Chicago, IL 60604
312.886.6283
312.353.8478 FAX

Region 6—(AR, LA, NM, OK, TX)
Technical Assistance
U.S. DOL/OSHA
525 Griffin Street - Room 602
Dallas, TX 75202
972.850.4180
972.850.4150 FAX

Technical Assistance
U.S. DOL/OSHA
606 N. Carancahua – Suite 700
Corpus Christi, TX 78476
361.888.3420 ext. 230
361.888.3424 FAX

U.S. DOL/OSHA
8713 Airport Freeway – Suite 302
Fort Worth, TX 76180
817.428.2470 ext. 230
817.581.7723 FAX

Region 7—(IA, KS, MO, NE)
Technical Assistance
U.S. DOL/OSHA
1100 Main Street - Suite 800
Center City Square
Kansas City, MO 64105
816.426.5861
816.426.2750 FAX

Region 8—(CO, MT, ND, SD, UT, WY)
Technical Assistance
U.S. DOL/OSHA
1999 Broadway - Suite 1690
Denver, CO 80202-5716
720.264.6558
303.844.1616 FAX

Region 9—(American Samoa, AZ, CA, Guam, HI, NV, Trust Territories of the Pacific)
Technical Assistance
U.S. DOL/OSHA
71 Stevenson Street, Room 420
San Francisco, CA 94105
800.475.4019
415.975.4319 FAX

Region 10—(AK, ID, OR, WA)
Technical Assistance
U.S. DOL/OSHA
1111 Third Avenue - Suite 715
Seattle, WA 98101-3212
206.553.5930
206.553.6499

State Plan States

The following states and territories operate their own OSHA-approved job safety and health programs. Connecticut and New York plans cover public employees only. States with approved programs must have a standard that is identical to, or at least as effective as, the Federal Standard. These state plan states are:

Alaska	Michigan	Puerto Rico
Arizona	Minnesota	South Carolina
California	Nevada	Tennessee
Connecticut	New Jersey	Utah
Hawaii	New Mexico	Vermont
Indiana	New York	Virgin Islands
Kentucky	North Carolina	Virginia
Maryland	Oregon	Wyoming

Assistance Services Provided by OSHA

Source: OSHA

OSHA can provide extensive help through a variety of programs, including assistance about safety and health programs, state plans, workplace consultations, voluntary protection programs, strategic partnerships, training and education, and more.

Consultation services

Consultation assistance is available on request to employers who want help in establishing and maintaining a safe and healthful workplace. Funded largely by OSHA, the service is provided at no cost to the employer. Primarily developed for smaller employers with more hazardous operations, the consultation service is delivered by state governments employing professional safety and health consultants. Comprehensive assistance includes a hazard survey of the worksite and appraisal of all aspects of the employer's existing safety and health management system. In addition, the service offers assistance to employers in developing and implementing an effective safety and health management system. No penalties are proposed or citations issued for hazards identified by the consultant. The employer's only obligation is to correct all identified serious hazards within the agreed upon correction timeframe. OSHA provides consultation assistance to the employer with the assurance that his or her name and firm and any information about the workplace will not be routinely reported to OSHA enforcement staff.

Under the consultation program, certain exemplary employers may request participation in OSHA's Safety and Health Achievement Recognition Program (SHARP). Eligibility for participation in SHARP includes, but is not limited to, receiving a full-service, comprehensive consultation visit, correcting all identified hazards, and developing an effective safety and health program management system.

Employers accepted into SHARP may receive an exemption from programmed inspections (not complaint or accident investigation inspections) for a period of 1 year initially, or 2 years upon renewal. For more information concerning consultation assistance, see the list of consultation projects listed at the end of this publication.

Voluntary Protection Programs (VPP)

Voluntary Protection Programs and onsite consultation services, when coupled with an effective enforcement program, expand worker protection to help meet the goals of the OSH Act. The three levels of VPP—Star, Merit, and Demonstration—are designed to recognize outstanding achievements by companies that have developed and implemented effective safety and health management systems. The VPPs motivate others to achieve excellent safety and health results in the same outstanding way as they establish a cooperative relationship between employers, employees and OSHA.

For additional information on VPPs and how to apply, contact the nearest OSHA regional office.

Strategic Partnership Program

OSHA's Strategic Partnership Program, the newest member of OSHA's cooperative programs, helps encourage, assist, and recognize the efforts of partners to eliminate serious workplace hazards and achieve a high level of worker safety and health. Whereas OSHA's Consultation Program and VPP entail one-on-one relationships between OSHA and individual worksites, most strategic partnerships seek to have a broader impact by building cooperative relationships with groups of employers and employees. These partnerships are voluntary, cooperative relationships between OSHA, employers, employee representatives, and others such as trade unions, trade and professional associations, universities, and other government agencies.

For more information on this program, contact the nearest OSHA office, or visit OSHA's website at www.osha.gov.

Training and Education

OSHA's area offices offer a variety of information services, such as compliance assistance, technical advice, publications, audiovisual aids, and speakers for special engagements. OSHA's Training Institute in Des Plaines, IL, provides basic and advanced courses in safety and health for federal and state compliance officers, state consultants, federal agency personnel, and private sector employers, employees, and their representatives.

The OSHA Training Institute also has established OSHA Training Institute Education Centers to address the increased demand for its courses from the private sector and from other federal agencies. These centers are nonprofit colleges, universities, and other organizations that have been selected after a competition for participation in the program.

OSHA awards grants through its Susan Harwood Training Grant Program to nonprofit organizations to provide safety and health training and education to employers and workers in the workplace. The grants focus on programs that will educate workers and employers in small business (fewer than 250 employees), training workers and employers about new OSHA standards or about high-risk activities or hazards. Grants are awarded for one year and may be renewed for an additional 12-month period depending on whether the grantee has performed satisfactorily.

OSHA expects each organization awarded a grant to develop a training and/or education program that addresses a safety and health topic named by OSHA, recruit workers and employers for the training, and conduct the training. Grantees are also expected to follow up with people who have been trained to find out what changes were made to reduce the hazards in their workplaces as a result of the training.

Each year OSHA has a national competition that is announced in the Federal Register and on the Internet at www.osha.gov/Training/sharwood/sharwood.html. For more information on grants, training, and education, contact the OSHA Training Institute, Office of Training and Education, 1555 Times Drive, Des Plaines, IL 60018; call (847) 297-4810, or see Outreach on OSHA's website at *www.osha.gov*.

Electronic Information

OSHA has a variety of materials and tools available on its website at www.osha.gov. These include e-Tools such as Expert Advisors and Electronic Compliance Assistance Tools (e-CATs), Technical Links, regulations, directives, publications, videos, and other information for employers and employees. OSHA's software programs and compliance assistance tools "walk" the employer/employee through challenging safety and health issues and common problems to find the best solutions for the workplace.

OSHA's CD-ROM includes standards, interpretations, directives, and more and can be purchased on CD-ROM from the U.S. Government Printing Office. To order, write to the Superintendent of Documents, U.S. Government Printing Office, Washington, DC 20402, or phone (202) 512-1800.

OSHA Publications

OSHA has an extensive publications program. For a listing of free or sales items, visit OSHA's website at www.osha.gov or contact the OSHA Publications Office, U.S. Department of Labor, OSHA/OSHA Publications, 200 Constitution Avenue, NW, N-3101, Washington, DC 20013-7535. Telephone (202) 693-1888 or FAX to (202) 693-2498.

Emergencies, Complaints, and Further Assistance

To report an emergency, file a complaint, or seek OSHA advice, assistance or products call (800) 321-OSHA or contact your nearest OSHA regional, area, state plan, or consultation office. The TTY number is (877) 889-5627.

A complaint may also be filed online. More information on federal and state programs may be obtained by visiting OSHA's website at www.osha.gov.

Appeals in States and Territories with OSHA-approved Plans

States with their own occupational safety and health programs have a state system for review and appeal of citations, penalties, and abatement periods. The procedures are generally similar to Federal OSHA's, but a state review board or equivalent authority hears cases.

Directory of Approved Sources of OSHA Funded Consultation

ALABAMA
Safe State Program
University of Alabama
432 Martha Parham West
Box 870388
Tuscaloosa, Alabama 35487
205.348.3033
205.348.3049 (FAX)

ALASKA
Consultation Section, ADOL/AKOSH
3301 Eagle Street, Suite 305
Anchorage, Alaska 99503
907.269.4957
907.2694950 (FAX)

ARIZONA
Consultation & Training
Industrial Commission of Arizona
Division of Occupational Safety & Health
2675 East Broadway Road; Suite 239
Tucson, Arizona 85716
520.628.5478
520.322.8008 (FAX)

ARKANSAS
OSHA Consultation
Arkansas Department of Labor
10421 West Markham
Little Rock, Arkansas 72205
501.682.4522
501.682.4532 (FAX)

CALIFORNIA
CAL/OSHA Consultation Service
Department of Industrial Relations

2424 Arden Way, Suite 485
Sacramento, California 95825
916.263.5765
916.263.5768 (FAX)

COLORADO
Colorado State University
Department of Environmental and Radiological Health Services
Occupational Health and Safety Consultation
1681 Campus Delivery
Fort Collins, Colorado 80523
970.491.6151
970.491.7778 (FAX)

CONNECTICUT
Connecticut Department of Labor
Division of Occupational Safety & Health
38 Wolcott Hill Road
Wethersfield, Connecticut 06109
860.263.6900
860.263.6940 (FAX)

DELAWARE
Delaware Department of Labor
Occupational Safety & Health
Division of Industrial Affairs
4425 North Market Street
Wilmington, Delaware 19802
302.761.8219
302.761.6602 (FAX)

DISTRICT OF COLUMBIA
Office of Occupational Safety & Health
D.C. Department of Employment Services
64 New York Avenue, NE-Room 2106
Washington, D.C. 20002
202.671.1800
202.673.2380 (FAX)
(Program available only for employers within the District of Columbia)

FLORIDA
Safety Florida Consultation Program
University of South Florida
13201 Bruce B. Downs Boulevard, MDC 56
Tampa, Florida 33612
813.974.9971
813.974.8270 (FAX)

GEORGIA
21(d) Onsite Consultation Program
430 10th St., N.E.
North Building
Atlanta, Georgia 30332-0837
404.407.8276
404.894.8275 (FAX)

GUAM
Guam Department of Labor
OSHA Onsite Consultation Program
GCIC Building, 8th Floor
414 West Soledad Avenue
Hagatna, Guam 96910
671.475.7069
671.475.7070 (FAX)

HAWAII
Consultation & Training Branch
Department of Labor & Industrial Relations
830 Punchbowl Street
Room #425
Honolulu, Hawaii 96813
808.586.9100
808.586.9104 (FAX)

IDAHO
Idaho OSHA Consultation Program
Boise State University
1910 University Drive, M.S. 1825
Boise, Idaho 83725-1825

208.426.3283
208.426.4411 (FAX)

ILLINOIS
Illinois Onsite Consultation
Industrial Service Division
Department of Commerce & Economic Opportunity
State of Illinois Center, Suite 3-400
100 West Randolph Street
Chicago, Illinois 60601
312.814.2337
312.814.7238 (FAX)

INDIANA
INSafe
Indiana Department of Labor
402 West Washington, W195
Indianapolis, Indiana 46204
317.232.2688
317.232.1868 FAX

IOWA
Iowa Workforce Development
Division of Labor Services
Bureau of Consultation and Education
1000 East Grand Avenue
Des Moines, Iowa 50319
515.281.7629
515.281.5522 (FAX)

KANSAS
Kansas Consultation Project
Kansas Department of Labor
800 SW Jackson – Suite 1500
Topeka, Kansas 66612-1227
785.296.6325
785.296.1775 (FAX)

KENTUCKY
Kentucky Labor Cabinet

Division of Education & Training
Kentucky OSH Program
1047 U.S. Highway 127, South, Suite 4
Frankfort, Kentucky 40601
502.564.3070
502.564.4769 (FAX)

LOUISIANA
21(d) Consultation Program
Louisiana Department of Labor
1001 N. 23rd Street, Room 421
P.O. Box 94040
Baton Rouge, Louisiana 70804-9094
225.342.9601
225.342.5158 (FAX)

MAINE
Maine Bureau of Labor Standards
Workplace Safety & Health Division
State House Station #45
Augusta, Maine 04333-0045
207.624.6463
207.624.6449 (FAX)

MARYLAND
MOSH Consultation Services
Montgomery Park Business Center
1827 Washington Blvd.
Baltimore, MD 21230
410.537.4500
410.537.4518 (FAX)

MASSACHUSETTS
Division of Occupational Safety
Dept. of Labor
1001 Watertown Street
West Newton, Massachusetts 02465
617.969.7177
617.727.4581 (FAX)

MICHIGAN
Consultation Education & Training Division
Michigan Occupational Safety and Health Administration
Labor & Economic Growth
7150 Harris Drive
P.O. Box 30643
Lansing, Michigan 48909-8413
517.322.1809
517.322.1374 (FAX)

MINNESOTA
Department of Labor & Industry
Consultation Division
443 LaFayette Road
Saint Paul, Minnesota 55155
651.284.5060
651.284.5393 (FAX)

MISSISSIPPI
Mississippi State University
Center for Safety and Health
2151 Hwy 18
Suite B
Brandon, MS 39042
601.825.0783
601.825.6609 (FAX)

MISSOURI
Onsite Consultation Program
Division of Labor Standards
Dept. of Labor & Industrial Relations
3315 West Truman Boulevard
Post Office Box 449
Jefferson City, Missouri 65102
573.751.3403
573.751.3721 (FAX)

MONTANA
Department of Labor & Industry
Occupational Safety and Health Bureau

PO Box 1728
1625 11th Avenue
Helena, Montana 59601
406.444.6401
406.444.9396 (FAX)

NEBRASKA
Nebraska Workforce Development
Office of Safety & Labor Standards
State Office Building, Lower Level
Post Office Box 95024
301 Centennial Mall, South
Lincoln, Nebraska 68509-5024
402.471.4717
402.471.5039 (FAX)

NEVADA
Safety Consultation & Training Section
Division of Industrial Relations
Department of Business & Industry
1301 North Green Valley Parkway #200
Henderson, Nevada 89074
702.486.9159
702.990.0362 (FAX)

NEW HAMPSHIRE
Occupational Health and Safety Consultation Service
New Hampshire Department of Environmental Services
P.O. Box 95
29 Hazen Drive
Concord, New Hampshire 03302-0095
603.271.2024
603.271.2667 (FAX)

NEW JERSEY
New Jersey Department of Labor
1 John Finch Plaza
P.O. Box 386
Trenton, New Jersey 08625-0386

609.777.0249
609.695.1314 (FAX)

NEW MEXICO
New Mexico Environment Department
Occupational Health & Safety Bureau
525 Camino DeLos Marquez, Suite 3
PO Box 26110
Santa Fe, New Mexico 87502
505.476.8700
505.476.8734 (FAX)
877.610.6742 (Toll Free)

NEW YORK
Division of Safety & Health
State Office Building Campus
Building 12, Room 168
Albany, New York 12240
518.457.2238
518.457.3454 (FAX)

NORTH CAROLINA
Bureau of Consultative Services
North Carolina Department of Labor
1101 Mail Service Center
Raleigh, North Carolina 27699-1101
919.807.2905
919.807.2902 (FAX)

NORTH DAKOTA
North Dakota Occupational Safety & Health
Consultation - Bismarck State College
Corporate & Continuing Education
1815 Schafer Street
Bismarck, North Dakota 58501
701.224.5778
701.224.5763 (FAX)

NORTHERN MARIANA ISLANDS
CNMI Department of Labor

CNMI-OSHA Onsite Consultation Program
Afetnas Building, 2nd Floor, San Antonio
Saipan, MP 96950
670.236.0913
670.664.3158 (FAX)

OHIO
Ohio Bureau of Workers' Compensation
Division of Safety and Hygiene
OSHA On-Site Consultation
13430 Yarmouth Drive
Pickerington, Ohio 43147
800.282.1425 (Toll Free)
614.644.3133 (FAX)

OKLAHOMA
Oklahoma Department of Labor
OSHA Consultation Division
4001 North Lincoln Blvd.
Oklahoma City, Oklahoma 73105-5212
405.528.1500
405.557.1214 (FAX)

OREGON
Oregon OSHA Consultation and Services
Department of Consumer & Business Services
350 Winter Street, N.E., Room 430
Salem, Oregon 97301-3882
503.378.3272
503.947.7462 (FAX)

PENNSYLVANIA
Indiana University Pennsylvania
Room 210 Walsh Hall
302 East Walk
Indiana, Pennsylvania 15705-1087
800.3821241 (Toll Free)
724.357.2396
724.357.2385 (FAX)

PUERTO RICO
Puerto Rico Occupational Safety & Health Administration (PR OSHA)
Voluntary Programs Division
P.O. Box 195540
San Juan, Puerto Rico 00919-5540
800.981.5720
787.767.6051 (FAX)

RHODE ISLAND
OSH Consultation Program
Division of Occupational Health & Radiation Control
Rhode Island Department of Health
3 Capital Hill,
Cannon Building, Room 206
Providence, Rhode Island 02908
401.222.2438
401.222.2456 (FAX)

SOUTH CAROLINA
South Carolina Department of Labor, Licensing & Regulation
110 Centerview Dr.
P.O. Box 11329
Columbia, South Carolina 29211-1329
803.896.7744
803.896.7750 (FAX)

SOUTH DAKOTA
South Dakota State University
Engineering Extension
West Hall 118, Box 510
907 Harvey Dunn Street
Brookings, South Dakota 57007-0597
605.688.4101
605.688.6290 (FAX)

TENNESSEE
TOSHA Consultation Services
Tennessee Department of Labor and Workforce Development
3rd floor, Andrew Johnson Tower

710 James Robertson Parkway
Nashville, Tennessee 37243-0659
615.741.7155
615.532.2997 (FAX)

TEXAS
Workers' Health & Safety Division- MS 22
Texas Department of Insurance
7551 Metro Center Drive
Austin, Texas 78744-1609
512.804.4640
512.804.4641 (FAX)
800.687.7080 (OSHCON Request Line)

UTAH
Utah OSHA Consultation Program
160 East 300 South, Third Floor
Salt Lake City, Utah 84114-6650
801.530.6855
801.530.6992 (FAX)
801.530.6901 (TTY)
800.530.5090

VERMONT
Vermont Department of Labor & Industry
Division of Occupational Safety and Health
National Life Building, Drawer 20
Montpellier, VT 05602-3401
802.888.0620
802.828.2195 (FAX)

VIRGINIA
Virginia Department of Labor & Industry
Occupational Safety & Health
Training & Consultation
13 South 13th Street
Richmond, Virginia 23219
804.786.6613
804.786.8418 (FAX)

VIRGIN ISLANDS
Safety in Paradise
University of the Virgin Islands
Community Engagement and Lifelong Learning Center
#2 John Brewer's Bay
St. Thomas, Virgin Islands 00803
340.693.1100
340.693.1115 (FAX)

WASHINGTON
Washington Dept of Labor & Industries
WISHA Services Division
7273 Linderson Way, SW
Tumwater, Washington 98501-5414
360.902.5443
360.902.5459 (FAX)

WEST VIRGINIA
West Virginia Department of Labor
Capitol Complex Building #6
1800 East Washington Street, Room B-749
Charleston, West Virginia 25305
304.558.7890
304.558.2415 (FAX)

WISCONSIN (Health)
WI OSHA Health Consultation Program
University of Wisconsin State Laboratory of Hygiene
Environmental Health Division
2601 Agriculture Drive
P.O. Box 7996
Madison, WI 53707-7996
608.226.5240
608.226.5249 (FAX)

WISCONSIN (Safety)
Wisconsin Department of Commerce
Division of Marketing, Advocacy and Technical Development
141 NW Barstow Street

Waukesha, Wisconsin 53188-3789
262.521.5198
800.947.0553 (Toll Free)
262.521.5369 (FAX)

WYOMING
Wyoming Workers' Safety
1510 East Pershing Blvd.
Cheyenne, Wyoming 82002
307.777.7786
307.777.3646 (FAX)

ADA Information Sources

*The following federal agencies and other organizations provide information about the **Americans with Disabilities Act (ADA)** and informal guidance in understanding and complying with different provisions of the ADA.*

The **Department of Justice** (DOJ) offers technical assistance on the ADA Standards for Accessible Design and other ADA provisions applying to businesses, non-profit service agencies, and state and local government programs. The DOJ also provides information on how to file ADA complaints.

ADA Information Line:
- 800.514.0301 (voice)
- 800.514.0383 (TTY)
- www.ada.org

The Equal Employment Opportunity Commission (EEOC) offers technical assistance on the ADA provisions applying to employment, also provides information on how to file ADA complaints.

Employment questions:
- 800.669.4000 (voice)
- 800.669.6820 (TTY)

Employment publications:

- 800.669.3362 (voice)
- 800.800.3302 (TTY)
- www.eeoc.gov

The Department of Transportation, Federal Transit Administration offers information on the public transit provisions of the ADA.

ADA Assistance Line:

- 888.446.4511 (voice)
- TTY: use relay service
- FTA.ADAassistance@dot.gov (email)
- www.fat.dot.gov/ada

The **Federal Communications Commission (FCC)** offers technical assistance on the ADA's Telephone Relay Service (TRS) requirements.

TRS publications and questions:

- 888.225.5322 (voice)
- 888.835.5322 (TTY)
- www.fcc.gov/cgb/dro

The Department of Education funds ten regional centers to provide technical assistance on the ADA.

ADA & IT Technical Assistance Centers:

- 800.949.4232 (voice/TTY)
- www.adata.org

The **Access Board (or Architectural and Transportation Barriers Compliance Board)** offers technical assistance on the ADA Accessibility Guidelines.

Publications and questions:

- 800.872.2253 (voice)
- 800.993.2822 (TTY)

- www.access-board.gov
- ta@access-board.gov

The **Job Accommodation Network (JAN)**, is funded by the Department of Labor to provide advice on accommodating employees with disabilities.

Job Accommodation Network:

- 800.526.7234 (voice)
- 877.781.9403 (TTY)
- www.jan.wvu.edu

The **Project ACTION** is funded by the Department of Transportation to provide information about making transportation accessible.

Transportation information and publications:

- 800.659.6428 (voice)
- TTY: use relay service
- http://projectaction.easterseals.com

The **Internal Revenue Service** offers tax incentives to help cover the cost of complying with the ADA. Form 8826 provides a tax credit for small businesses. Publication 535 explains tax deductible expenses for business of any size.

Tax forms and publications:

- 800.829.3676 (voice)
- 800.829.4059 (TTY)
- www.irs.gov

Appendix II
OSHA Chronology

OSHA's mission is to send every worker home whole and healthy every day. Since the agency was established in 1971, workplace fatalities have been cut by 62 percent and occupational injury and illness rates have declined 40 percent. At the same time, U.S. employment has nearly doubled from 56 million workers at 3.5 million worksites to 115 million workers at nearly 7 million sites. The following milestones mark the agency's progress over the past 35 years in improving working environments for America's workforce.

December 29, 1970	President Richard M. Nixon signed the Occupational Safety and Health Act of 1970.
May 29, 1971	First standards were adopted to provide baseline for safety and health protection in American workplaces.
January 17, 1972	OSHA Training Institute was established to instruct OSHA inspectors and the public.
November-December, 1972	South Carolina, Montana and Oregon were the first states approved to run their own OSHA programs.
May 20, 1975	Free consultation program was created; more than 500,000 businesses participated in the past 30 years.
June 23, 1978	Cotton dust standard was promulgated to protect 600,000 workers from byssinosis (brown lung disease). The cases of "brown lung" have declined to 0.1 cases per 10,000 workers.

January 20, 1978	Supreme Court decision set staffing benchmarks for state plans to be "at least as effective" as federal OSHA.
April 12, 1978	New Directions Grants program, now known as the Susan Harwood Training Grants program, was created to foster development of occupational safety and health training and education for employers and workers. More than 1.3 million participants were trained since 1978.
November 14, 1978	Lead standard was published in order to reduce permissible exposures by three-quarters thereby protecting 835,000 workers from damage to nervous, urinary and reproductive systems. The construction standard adopted in 1995.
February 26, 1980	Supreme Court decision on Whirlpool affirmed workers' rights to engage in safety and health-related activities.
May 23, 1980	Medical and exposure records standard was finalized to permit worker and OSHA access to employer-maintained medical and toxic exposure records.
July 2, 1980	Supreme Court decision vacated OSHA's benzene standard, thereby establishing the principle that OSHA standards must address and reduce "significant risks" to workers.
September 12,1980	Fire protection standard was updated and rules were established for fire brigades responsible for putting out nearly 95 percent of worksite fires.
January 16, 1981	Electrical standards were updated to simplify compliance and adopt a performance approach.
July 2, 1982	Voluntary Protection Programs (VPP) were cre-

	ated to recognize worksites with outstanding safety and health programs; more than 1,400 sites are currently participating.
November 25,1983	Hazard communication standard was promulgated to provide information and training and labeling of toxic materials for manufacturing employers and employees. Other industries were added August 24, 1987.
November-December, 1984	First "final approvals" were granted to the state plans of Virgin Islands, Hawaii and Alaska giving them authority to operate with minimal oversight from OSHA.
April 1, 1986	First instance-by instance penalties were proposed against Union Carbide's plant in Institute, West Virginia, for egregious violations involving respiratory protection and injury and illness recordkeeping.
December 31, 1987	Grain handling facilities standard was adopted to protect 155,000 workers at nearly 24,000 grain elevators from the risk of fire and explosion from highly combustible grain dust.
January 26, 1989	*"Safety and Health Program Management Guidelines,"* the voluntary guidelines for effective safety and health programs based on the VPP (Voluntary Protection Program) experience, were published.
March 6, 1989	Hazardous waste operations and emergency response standard was promulgated to protect 1.75 million public and private sector workers exposed to toxic wastes from spills or at hazardous waste sites.
September 1,1989	Lockout/tagout of hazardous energy sources

	standard was issued in order to protect 39 million workers from the unexpected start up of machines or equipment. The standard was also issued to prevent 120 deaths and 50,000 injuries each year.
December 6, 1991	Occupational exposure to bloodborne pathogens standard was published to prevent more than 9,000 infections and 200 deaths per year, protecting 5.6 million workers against AIDS, hepatitis B and other diseases.
October 1, 1992	Education Centers were created to make OSHA training courses more widely available to employers, workers and the public. Twenty centers train more than 300,000 students each year; over 370,000 students were trained in 2005 alone.
February 24,1992	Process safety management of highly hazardous chemicals standard was adopted in order to reduce fire and explosion risks for 3 million workers at 25,000 workplaces; preventing more than 250 deaths and more than 1,500 injuries each year.
January 14,1993	Permit-required confined spaces standard was promulgated to prevent more than 50 deaths and more than 5,000 serious injuries annually for the 1.6 million workers who enter confined spaces at 240,000 workplaces each year.
February 1, 1993	Maine 200 program was created to promote development of safety and health programs at companies with high numbers of injuries and illnesses.
June 27, 1994	First expert advisor software, *GoCad,* was issued in order to assist employers in complying with OSHA's cadmium standard.

August 9,1994	Fall protection in construction standard was revised to save 79 lives and prevent 56,400 injuries each year.
August 10,1994	Asbestos standard was updated to cut permissible exposures in half for nearly 4 million workers, preventing 42 cancer deaths annually.
September 4,1995	Formal launch of OSHA's expanded webpage provides OSHA standards and compliance assistance via the Internet.
June 6, 1996	Phone/fax complaint handling policy was adopted in order to speed resolution of complaints of unsafe or unhealthful working conditions.
August 30,1996	Scaffold standard was published to protect 2.3 million construction workers and prevent 50 deaths and 4,500 injuries annually.
November 9,1998	OSHA Strategic Partnership Program was launched to improve workplace safety and health through national and local cooperative, voluntary agreements.
April 19,1999	Site-Specific Targeting Program was established to focus OSHA resources where most needed—on individual worksites with the highest injury and illness rates.
November 14, 2000	Ergonomics program standard was promulgated to prevent 460,000 musculoskeletal disorders among the more than 102 million workers at 6.1 million general industry worksites.
January 10, 2001	New Jersey public employee plan receives final approval.
January 17, 2001	Steel erection standard was issued. It was devel-

	oped in concert with industry and union groups, thereby preventing 30 fatalities and 1,142 injuries annually and saving employers nearly $40 million a year. It is the first OSHA safety standard to be developed under the negotiated rulemaking process.
January 18, 2001	Recordkeeping rule was revised to improve the system that employers use to track and record workplace injuries and illnesses.
January 18, 2001	As mandated by the Needlestick Safety and Prevention Act, OSHA revised its bloodborne pathogens standard to clarify the need for employers to select safer needle devices.
March 7-8, 2001	Under the auspices of the Congressional Review Act, the Senate voted 56-44 to repeal ergonomics rule. The House followed suit the next day and voted 223-206 to repeal the rule. It is the first time that Congress exercised its authority under the Act to repeal a federal standard.
March 20, 2001	President George W. Bush signed S.J. Resolution 6, repealing the ergonomics rule.
April 27, 2001	Occupational Safety and Health Administration celebrates its 30th anniversary. Over the past three decades job-related fatalities are cut in half; with injuries and illnesses declining by 40 percent.
September 11, 2001	OSHA responded to the terrorist attacks at the World Trade Center in New York City and the Pentagon outside Washington, DC. More than 1,000 OSHA employees from New York and around the country volunteered to help protect workers involved in the cleanup and recovery efforts at both sites.

March 1, 2002	Agency launches its bi-weekly electronic newsletter *QuickTakes*.
April 4, 2002	Secretary Chao unveiled a comprehensive plan designed to reduce ergonomic injuries through a combination of industry-specific guidelines, strong enforcement, outreach and assistance, and further research.
May 30, 2002	The recovery phase and cleanup at the World Trade Center disaster site comes to an end. For more than eight months, three million work hours were logged on a worksite like no other, yet only 35 workers missed workdays due to injury and no more lives were lost to work.
March 11, 2003	OSHA announced an enhanced enforcement policy to focus on those employers who have received "high gravity" citations.
March 13, 2003	Ergonomics Guidelines were issued for the Nursing Home Industry.
July 1, 2003	Final rule established the criteria for recording work-related hearing loss.
October 24, 2003	OSHA welcomed 1000th site to achieve "Star" status in Voluntary Protection Program (VPP).
February 4, 2004	OSHA unveiled its National Emergency Management Plan, a new directive that clarifies the agency's policies during responses to national emergencies.
May 28, 2004	Ergonomics Guidelines were published for Retail Grocery Stores.
August 24, 2004	Final rule established procedures for handling whistleblower complaints under the Corporate

	and Criminal Fraud Accountability Act of 2002, also known as the Sarbanes-Oxley Act.
September 2, 2004	Ergonomic guidelines were announced for the poultry processing industry.
November 24, 2004	Federal agencies were required to adopt private sector worker safety and health recordkeeping and reporting requirements.
May 12, 2005	Oregon became the 17th state to receive final approval to operate their own job safety and health program.

Source: Occupational Safety and Health Administration

Appendix III

Self-Inspection Checklists

Source: Occupational Safety And Health Administration, U.S. Department of Labor

SELF-INSPECTION

The most widely accepted way to identify hazards is to conduct safety and health inspections because the only way to be certain of an actual situation is to look at it directly from time to time.

Begin a program of self-inspection in your own workplace. Self-inspection is essential if you are to know where probable hazards exist and whether they are under control.

This section includes checklists designed to assist you in self-inspection fact-finding. The checklists can give you some indication of where to begin taking action to make your business safer and more healthful for all of your employees.

These checklists are by no means all-inclusive and not all of the checklists will apply to your business. You might want to start by selecting the areas that are most critical to your business, then expanding your self-inspection checklists over time to fully cover all areas that pertain to your business. Remember that a checklist is a tool to help, not a definitive statement of what is mandatory. Use checklists only for guidance.

Don't spend time with items that have no application to your business. Make sure that each item is seen by you or your designee and leave nothing to memory or chance. Write down what you see or don't see and what you think you should do about it.

Add information from your completed checklists to injury information, employee information, and process and equipment information to build a foundation to help you determine what problems exist. Then, as you use the OSHA standards in your problem-solving process, it will be easier for you to determine the actions needed to solve these problems.

Once the hazards have been identified, institute the control procedures described at page 9 and establish your four-point safety and health program.

Self-Inspection Scope

Your self-inspections should cover safety and health issues in the following areas:

- **Processing, Receiving, Shipping and Storage** – equipment, job planning, layout, heights, floor loads, projection of materials, material handling and storage methods, training for material handling equipment.

- **Building and Grounds Conditions** – floors, walls, ceilings, exits, stairs, walkways, ramps, platforms, driveways, aisles.

- **Housekeeping Program** – waste disposal, tools, objects, materials, leakage and spillage, cleaning methods, schedules, work areas, remote areas, storage areas.

- **Electricity** – equipment, switches, breakers, fuses, switch-boxes, junctions, special fixtures, circuits, insulation, extensions, tools, motors, grounding, national electric code compliance.

- **Lighting** – type, intensity, controls, conditions, diffusion, location, glare and shadow control.

- **Heating and Ventilation** – type, effectiveness, temperature, humidity, controls, natural and artificial ventilation and exhausting.

- **Machinery** – points of operation, flywheels, gears, shafts, pulleys, key ways, belts, couplings, sprockets, chains, frames, controls, lighting for tools and equipment, brakes, exhausting, feeding, oiling, adjusting, maintenance, lockout/tagout, grounding, work space, location, purchasing standards.

- **Personnel** – training, including hazard identification training; experience; methods of checking machines before use; type of clothing; PPE; use of guards; tool storage; work practices; methods for cleaning, oiling, or adjusting machinery.

- **Hand and Power Tools** – purchasing standards, inspection, storage, repair, types, maintenance, grounding, use and handling.

- **Chemicals** – storage, handling, transportation, spills, disposals, amounts used, labeling, toxicity or other harmful effects, warning signs, supervision, training, protective clothing and equipment, hazard communication requirements.

- **Fire Prevention** – extinguishers, alarms, sprinklers, smoking rules, exits, personnel assigned, separation of flammable materials and dangerous operations, explosion-proof fixtures in hazardous locations, waste disposal and training of personnel.

- **Maintenance** – provide regular and preventive maintenance on all equipment used at the worksite, recording all work performed on the machinery and by training personnel on the proper care and servicing of the equipment.

- **PPE** – type, size, maintenance, repair, age, storage, assignment of responsibility, purchasing methods, standards observed, training in care and use, rules of use, method of assignment.

- **Transportation** – motor vehicle safety, seat belts, vehicle maintenance, safe driver programs.

- **First Aid Program/Supplies** – medical care facilities locations, posted emergency phone numbers, accessible first aid kits.

- **Evacuation Plan** – establish and practice procedures for an emergency evacuation, e.g., fire, chemical/biological incidents, bomb threat; include escape procedures and routes, critical plant operations, employee accounting following an evacuation, rescue and medical duties and ways to report emergencies.

Self-Inspection Checklists

These checklists are by no means all-inclusive. You should add to them or delete items that do not apply to your business; however, carefully consider each item and then make your decision. You should refer to OSHA standards for specific guidance that may apply to your work situation. (**Note:** These checklists are typical for general industry but not for construction or maritime industries.)

EMPLOYER POSTING

- ☐ Is the required OSHA Job Safety and Health Protection Poster displayed in a prominent location where all employees are likely to see it?

- ☐ Are emergency telephone numbers posted where they can be readily found in case of emergency?

- ☐ Where employees may be exposed to toxic substances or harmful physical agents, has appropriate information concerning employee access to medical and exposure records and Material Safety Data Sheets (MSDSs) been posted or otherwise made readily available to affected employees?

- ☐ Are signs concerning exit routes, room capacities, floor loading, biohazards, exposures to x-ray, microwave, or other harmful radiation or substances posted where appropriate?

- ☐ Is the Summary of Work-Related Injuries and Illnesses (OSHA Form 300A) posted during the months of February, March and April?

RECORDKEEPING

- ☐ Are occupational injuries or illnesses, except minor injuries requiring only first aid, recorded as required on the OSHA 300 log?

- ☐ Are employee medical records and records of employee exposure to hazardous substances or harmful physical agents up-to-date and in compliance with current OSHA standards?

- ☐ Are employee training records kept and accessible for review by employees, as required by OSHA standards?

- ☐ Have arrangements been made to retain records for the time period required for each specific type of record? (Some records must be maintained for at least 40 years.)

- ☐ Are operating permits and records up-to-date for items such as elevators, air pressure tanks, liquefied petroleum gas tanks, etc.?

SAFETY AND HEALTH PROGRAM

- ☐ Do you have an active safety and health program in operation that includes general safety and health program elements as well as the management of hazards specific to your worksite?

- ☐ Is one person clearly responsible for the safety and health program?

- ☐ Do you have a safety committee or group made up of management and labor representatives that meets regularly and reports in writing on its activities?

- ☐ Do you have a working procedure to handle in-house employee complaints regarding safety and health?
- ☐ Are your employees advised of efforts and accomplishments of the safety and health program made to ensure they will have a workplace that is safe and healthful?
- ☐ Have you considered incentives for employees or workgroups who excel in reducing workplace injury/illnesses?

MEDICAL SERVICES AND FIRST AID

- ☐ Is there a hospital, clinic, or infirmary for medical care near your workplace or is at least one employee on each shift currently qualified to render first aid?
- ☐ Have all employees who are expected to respond to medical emergencies as part of their job responsibilities received first aid training; had hepatitis B vaccination made available to them; had appropriate training on procedures to protect them from bloodborne pathogens, including universal precautions; and have available and understand how to use appropriate PPE to protect against exposure to bloodborne diseases?*

*Pursuant to an OSHA memorandum of July 1, 1992, employees who render first aid only as a collateral duty do not have to be offered pre-exposure hepatitis B vaccine only if the employer includes and implements the following requirements in his/her exposure control plan: (1) the employer must record all first aid incidents involving the presence of blood or other potentially infectious materials before the end of the work shift during which the first aid incident occurred; (2) the employer must comply with post-exposure evaluation, prophylaxis and follow-up requirements of the Bloodborne Pathogens standard with respect to "exposure incidents," as defined by the standard; (3) the employer must train designated first aid providers about the reporting procedure; (4) the employer must offer to initiate the hepatitis B vaccination series within 24 hours to all unvaccinated first aid providers who have rendered assistance in any situation involving the presence of blood or other potentially infectious materials.

- ☐ If employees have had an exposure incident involving bloodborne pathogens, was an immediate post-exposure medical evaluation and follow-up provided?
- ☐ Are medical personnel readily available for advice and consultation on matters of employees' health?
- ☐ Are emergency phone numbers posted?
- ☐ Are fully supplied first aid kits easily accessible to each work area, periodically inspected and replenished as needed?
- ☐ Have first aid kits and supplies been approved by a physician, indicating that they are adequate for a particular area or operation?
- ☐ Is there an eye-wash station or sink available for quick drenching or flushing of the eyes and body in areas where corrosive liquids or materials are handled?

FIRE PROTECTION

- ☐ Is your local fire department familiar with your facility, its location and specific hazards?
- ☐ If you have a fire alarm system, is it certified as required and tested annually?
- ☐ If you have interior standpipes and valves, are they inspected regularly?
- ☐ If you have outside private fire hydrants, are they flushed at least once a year and on a routine preventive maintenance schedule?
- ☐ Are fire doors and shutters in good operating condition?
- ☐ Are fire doors and shutters unobstructed and protected against obstructions, including their counterweights?
- ☐ Are fire door and shutter fusible links in place?
- ☐ Are automatic sprinkler system water control valves, air and water pressure checked periodically as required?
- ☐ Is the maintenance of automatic sprinkler systems assigned to responsible persons or to a sprinkler contractor?
- ☐ Are sprinkler heads protected by metal guards if exposed to potential physical damage?

- ☐ Is proper clearance maintained below sprinkler heads?
- ☐ Are portable fire extinguishers provided in adequate number and type and mounted in readily accessible locations?
- ☐ Are fire extinguishers recharged regularly with this noted on the inspection tag?
- ☐ Are employees periodically instructed in the use of fire extinguishers and fire protection procedures?

PERSONAL PROTECTIVE EQUIPMENT AND CLOTHING

- ☐ Has the employer determined whether hazards that require the use of PPE (e.g., head, eye, face, hand, or foot protection) are present or are likely to be present?
- ☐ If hazards or the likelihood of hazards are found, are employers selecting appropriate and properly fitted PPE suitable for protection from these hazards and ensuring that affected employees use it?
- ☐ Have both the employer and the employees been trained on PPE procedures, i.e., what PPE is necessary for job tasks, when workers need it, and how to properly wear and adjust it?
- ☐ Are protective goggles or face shields provided and worn where there is any danger of flying particles or corrosive materials?
- ☐ Are approved safety glasses required to be worn at all times in areas where there is a risk of eye injuries such as punctures, abrasions, contusions, or burns?
- ☐ Are employees who wear corrective lenses (glasses or contacts) in workplaces with harmful exposures required to wear *only* approved safety glasses, protective goggles, or use other medically approved precautionary procedures?
- ☐ Are protective gloves, aprons, shields, or other means provided and required where employees could be cut or where there is reasonably anticipated exposure to corrosive liquids, chemicals, blood, or other potentially infectious materials? See the OSHA Bloodborne Pathogens standard, 29 CFR 1910.1030(b), for the definition of "other potentially infectious materials."
- ☐ Are hard hats required, provided and worn where danger of falling objects exists?
- ☐ Are hard hats periodically inspected for damage to the shell and suspension system?
- ☐ Is appropriate foot protection required where there is the risk of foot injuries from hot, corrosive, or poisonous substances, falling objects, crushing, or penetrating actions?
- ☐ Are approved respirators provided when needed? (See 29 CFR 1910.134 for detailed information on respirators or check OSHA's website at www.osha.gov).
- ☐ Is all PPE maintained in a sanitary condition and ready for use?
- ☐ Are food or beverages consumed only in areas where there is no exposure to toxic material, blood, or other potentially infectious materials?
- ☐ Is protection against the effects of occupational noise provided when sound levels exceed those of the OSHA Noise standard?
- ☐ Are adequate work procedures, PPE and other equipment provided and used when cleaning up spilled hazardous materials?
- ☐ Are appropriate procedures in place to dispose of or decontaminate PPE contaminated with, or reasonably anticipated to be contaminated with, blood or other potentially infectious materials?

GENERAL WORK ENVIRONMENT

- ☐ Are all worksites clean, sanitary and orderly?
- ☐ Are work surfaces kept dry and appropriate means taken to assure the surfaces are slip-resistant?
- ☐ Are all spilled hazardous materials or liquids, including blood and other potentially infectious materials, cleaned up immediately and according to proper procedures?
- ☐ Is combustible scrap, debris and waste stored safely and removed from the worksite promptly?

- ☐ Is all regulated waste, as defined in the OSHA Bloodborne Pathogens standard (29 CFR 1910.1030), discarded according to Federal, state and local regulations?
- ☐ Are accumulations of combustible dust routinely removed from elevated surfaces including the overhead structure of buildings, etc.?
- ☐ Is combustible dust cleaned up with a vacuum system to prevent suspension of dust particles in the environment?
- ☐ Is metallic or conductive dust prevented from entering or accumulating on or around electrical enclosures or equipment?
- ☐ Are covered metal waste cans used for oily or paint-soaked waste?
- ☐ Are all oil and gas-fired devices equipped with flame failure controls to prevent flow of fuel if pilots or main burners are not working?
- ☐ Are paint spray booths, dip tanks, etc., cleaned regularly?
- ☐ Are the minimum number of toilets and washing facilities provided and maintained in a clean and sanitary fashion?
- ☐ Are all work areas adequately illuminated?
- ☐ Are pits and floor openings covered or otherwise guarded?
- ☐ Have all confined spaces been evaluated for compliance with 29 CFR 1910.146? (Permit-required confined spaces.)

WALKWAYS

- ☐ Are aisles and passageways kept clear and marked as appropriate?
- ☐ Are wet surfaces covered with non-slip materials?
- ☐ Are holes in the floor, sidewalk, or other walking surface repaired properly, covered, or otherwise made safe?
- ☐ Is there safe clearance for walking in aisles where motorized or mechanical handling equipment is operating?
- ☐ Are materials or equipment stored in such a way that sharp projections will not interfere with the walkway?
- ☐ Are spilled materials cleaned up immediately?
- ☐ Are changes of direction or elevations readily identifiable?
- ☐ Are aisles or walkways that pass near moving or operating machinery, welding operations, or similar operations arranged so employees will not be subjected to potential hazards?
- ☐ Is adequate headroom provided for the entire length of any aisle or walkway?
- ☐ Are standard guardrails provided wherever aisle or walkway surfaces are elevated more than 30 inches (76.20 centimeters) above any adjacent floor or the ground?
- ☐ Are bridges provided over conveyors and similar hazards?

FLOOR AND WALL OPENINGS

- ☐ Are floor openings guarded by a cover, a guardrail, or equivalent on all sides (except at stairways or ladder entrances)?
- ☐ Are toeboards installed around the edges of permanent floor openings where persons may pass below the opening?
- ☐ Are skylight screens able to withstand a load of at least 200 pounds (90.7 kilograms)?
- ☐ Is the glass in windows, doors, glass walls, etc., subject to possible human impact, of sufficient thickness and type for the condition of use?
- ☐ Are grates or similar type covers over floor openings such as floor drains designed to allow unimpeded foot traffic or rolling equipment?
- ☐ Are unused portions of service pits and pits not in use either covered or protected by guardrails or equivalent?
- ☐ Are manhole covers, trench covers and similar covers, and their supports designed to carry a truck rear axle load of at least 20,000 pounds (9,072 kilograms) when located in roadways and subject to vehicle traffic?
- ☐ Are floor or wall openings in fire-resistant construction provided with doors or covers compatible with the fire rating of the structure and

provided with a self-closing feature when appropriate?

STAIRS AND STAIRWAYS

- [ ] Do standard stair rails or handrails on all stairways have at least four risers?
- [ ] Are all stairways at least 22 inches (55.88 centimeters) wide?
- [ ] Do stairs have landing platforms not less than 30 inches (76.20 centimeters) in the direction of travel and extend 22 inches (55.88 centimeters) in width at every 12 feet (3.6576 meters) or less of vertical rise?
- [ ] Do stairs angle no more than 50 and no less than 30 degrees?
- [ ] Are stairs of hollow-pan type treads and landings filled to the top edge of the pan with solid material?
- [ ] Are step risers on stairs uniform from top to bottom?
- [ ] Are steps slip-resistant?
- [ ] Are stairway handrails located between 30 inches (76.20 centimeters) and 34 inches (86.36 centimeters) above the leading edge of stair treads?
- [ ] Do stairway handrails have at least 3 inches (7.62 centimeters) of clearance between the handrails and the wall or surface they are mounted on?
- [ ] Where doors or gates open directly on a stairway, is a platform provided so the swing of the door does not reduce the width of the platform to less than 21 inches (53.34 centimeters)?
- [ ] Are stairway handrails capable of withstanding a load of 200 pounds (90.7 kilograms), applied within 2 inches (5.08 centimeters) of the top edge in any downward or outward direction?
- [ ] Where stairs or stairways exit directly into any area where vehicles may be operated, are adequate barriers and warnings provided to prevent employees from stepping into the path of traffic?
- [ ] Do stairway landings have a dimension measured in the direction of travel at least equal to the width of the stairway?
- [ ] Is the vertical distance between stairway landings limited to 12 feet (3.6576 meters) or less?

ELEVATED SURFACES

- [ ] Are signs posted, when appropriate, showing the elevated surface load capacity?
- [ ] Are surfaces that are elevated more than 30 inches (76.20 centimeters) provided with standard guardrails?
- [ ] Are all elevated surfaces beneath which people or machinery could be exposed to falling objects provided with standard 4-inch (10.16-centimeter) toeboards?
- [ ] Is a permanent means of access and egress provided to elevated storage and work surfaces?
- [ ] Is required headroom provided where necessary?
- [ ] Is material on elevated surfaces piled, stacked, or racked in a manner to prevent it from tipping, falling, collapsing, rolling, or spreading?
- [ ] Are dock boards or bridge plates used when transferring materials between docks and trucks or railcars?

EXITING OR EGRESS - EVACUATION

- [ ] Are all exits marked with an exit sign and illuminated by a reliable light source?
- [ ] Are the directions to exits, when not immediately apparent, marked with visible signs?
- [ ] Are doors, passageways or stairways that are neither exits nor access to exits, but could be mistaken for exits, appropriately marked "NOT AN EXIT," "TO BASEMENT," "STOREROOM," etc.?
- [ ] Are exit signs labeled with the word "EXIT" in lettering at least 5 inches (12.70 centimeters) high and the stroke of the lettering at least l/2-inch (1.2700 centimeters) wide?
- [ ] Are exit doors side-hinged?

- [ ] Are all exits kept free of obstructions?
- [ ] Are at least two means of egress provided from elevated platforms, pits, or rooms where the absence of a second exit would increase the risk of injury from hot, poisonous, corrosive, suffocating, flammable, or explosive substances?
- [ ] Are there sufficient exits to permit prompt escape in case of emergency?
- [ ] Are special precautions taken to protect employees during construction and repair operations?
- [ ] Is the number of exits from each floor of a building and the number of exits from the building itself appropriate for the building occupancy load?
- [ ] Are exit stairways that are required to be separated from other parts of a building enclosed by at least 2-hour fire-resistive construction in buildings more than four stories in height, and not less than 1-hour fire-resistive construction elsewhere?
- [ ] Where ramps are used as part of required exiting from a building, is the ramp slope limited to 1 foot (0.3048 meter) vertical and 12 feet (3.6576 meters) horizontal?
- [ ] Where exiting will be through frameless glass doors, glass exit doors, storm doors, etc., are the doors fully tempered and meet the safety requirements for human impact?

EXIT DOORS

- [ ] Are doors that are required to serve as exits designed and constructed so that the path of exit travel is obvious and direct?
- [ ] Are windows that could be mistaken for exit doors made inaccessible by means of barriers or railings?
- [ ] Are exit doors able to be opened from the direction of exit travel without the use of a key or any special knowledge or effort when the building is occupied?
- [ ] Is a revolving, sliding, or overhead door prohibited from serving as a required exit door?
- [ ] Where panic hardware is installed on a required exit door, will it allow the door to open by applying a force of 15 pounds (6.80 kilograms) or less in the direction of the exit traffic?
- [ ] Are doors on cold storage rooms provided with an inside release mechanism that will release the latch and open the door even if the door is padlocked or otherwise locked on the outside?
- [ ] Where exit doors open directly onto any street, alley, or other area where vehicles may be operated, are adequate barriers and warnings provided to prevent employees from stepping into the path of traffic?
- [ ] Are doors that swing in both directions and are located between rooms where there is frequent traffic provided with viewing panels in each door?

PORTABLE LADDERS

- [ ] Are all ladders maintained in good condition, joints between steps and side rails tight, all hardware and fittings securely attached, and moveable parts operating freely without binding or undue play?
- [ ] Are non-slip safety feet provided on each metal or rung ladder, and are ladder rungs and steps free of grease and oil?
- [ ] Are employees prohibited from placing a ladder in front of doors opening toward the ladder unless the door is blocked open, locked, or guarded?
- [ ] Are employees prohibited from placing ladders on boxes, barrels, or other unstable bases to obtain additional height?
- [ ] Are employees required to face the ladder when ascending or descending?
- [ ] Are employees prohibited from using ladders that are broken, have missing steps, rungs, or cleats, broken side rails, or other faulty equipment?
- [ ] Are employees instructed not to use the top step of ordinary stepladders as a step?

- ☐ When portable rung ladders are used to gain access to elevated platforms, roofs, etc., does the ladder always extend at least 3 feet (0.9144 meters) above the elevated surface?
- ☐ Are employees required to secure the base of a portable rung or cleat type ladder to prevent slipping, or otherwise lash or hold it in place?
- ☐ Are portable metal ladders legibly marked with signs reading "CAUTION - Do Not Use Around Electrical Equipment" or equivalent wording?
- ☐ Are employees prohibited from using ladders as guys, braces, skids, gin poles, or for other than their intended purposes?
- ☐ Are employees instructed to only adjust extension ladders while standing at a base (not while standing on the ladder or from a position above the ladder)?
- ☐ Are metal ladders inspected for damage?
- ☐ Are the rungs of ladders uniformly spaced at 12 inches (30.48 centimeters) center to center?

HAND TOOLS AND EQUIPMENT

- ☐ Are all tools and equipment (both company and employee-owned) used at the workplace in good condition?
- ☐ Are hand tools, such as chisels, punches, etc., which develop mushroomed heads during use, reconditioned or replaced as necessary?
- ☐ Are broken or fractured handles on hammers, axes and similar equipment replaced promptly?
- ☐ Are worn or bent wrenches replaced?
- ☐ Are appropriate handles used on files and similar tools?
- ☐ Are employees aware of hazards caused by faulty or improperly used hand tools?
- ☐ Are appropriate safety glasses, face shields, etc., used while using hand tools or equipment that might produce flying materials or be subject to breakage?
- ☐ Are jacks checked periodically to ensure they are in good operating condition?
- ☐ Are tool handles wedged tightly into the heads of all tools?
- ☐ Are tool cutting edges kept sharp so the tool will move smoothly without binding or skipping?
- ☐ Are tools stored in a dry, secure location where they cannot be tampered with?
- ☐ Is eye and face protection used when driving hardened or tempered studs or nails?

PORTABLE (POWER OPERATED) TOOLS AND EQUIPMENT

- ☐ Are grinders, saws and similar equipment provided with appropriate safety guards?
- ☐ Are power tools used with proper shields, guards, or attachments, as recommended by the manufacturer?
- ☐ Are portable circular saws equipped with guards above and below the base shoe?
- ☐ Are circular saw guards checked to ensure that they are not wedged up, leaving the lower portion of the blade unguarded?
- ☐ Are rotating or moving parts of equipment guarded to prevent physical contact?
- ☐ Are all cord-connected, electrically operated tools and equipment effectively grounded or of the approved double insulated type?
- ☐ Are effective guards in place over belts, pulleys, chains and sprockets on equipment such as concrete mixers, air compressors, etc.?
- ☐ Are portable fans provided with full guards or screens having openings 1/2 inch (1.2700 centimeters) or less?
- ☐ Is hoisting equipment available and used for lifting heavy objects, and are hoist ratings and characteristics appropriate for the task?
- ☐ Are ground-fault circuit interrupters provided on all temporary electrical 15 and 20 ampere circuits used during periods of construction?
- ☐ Are pneumatic and hydraulic hoses on power-operated tools checked regularly for deterioration or damage?

ABRASIVE WHEEL EQUIPMENT GRINDERS

- ☐ Is the work rest used and kept adjusted to within 1/8 inch (0.3175 centimeter) of the wheel?

- [ ] Is the adjustable tongue on the top side of the grinder used and kept adjusted to within 1/4 inch (0.6350 centimeters) of the wheel?
- [ ] Do side guards cover the spindle, nut and flange and 75 percent of the wheel diameter?
- [ ] Are bench and pedestal grinders permanently mounted?
- [ ] Are goggles or face shields always worn when grinding?
- [ ] Is the maximum revolutions per minute (rpm) rating of each abrasive wheel compatible with the rpm rating of the grinder motor?
- [ ] Are fixed or permanently mounted grinders connected to their electrical supply system with metallic conduit or other permanent wiring method?
- [ ] Does each grinder have an individual on and off control switch?
- [ ] Is each electrically operated grinder effectively grounded?
- [ ] Are new abrasive wheels visually inspected and ring tested before they are mounted?
- [ ] Are dust collectors and powered exhausts provided on grinders used in operations that produce large amounts of dust?
- [ ] Are splash guards mounted on grinders that use coolant to prevent the coolant from reaching employees?
- [ ] Is cleanliness maintained around grinders?

POWER-ACTUATED TOOLS

- [ ] Are employees who operate power-actuated tools trained in their use and required to carry a valid operator's card?
- [ ] Is each power-actuated tool stored in its own locked container when not being used?
- [ ] Is a sign at least 7 inches (17.78 centimeters) by 10 inches (25.40 centimeters) with bold face type reading "POWER-ACTUATED TOOL IN USE" conspicuously posted when the tool is being used?
- [ ] Are power-actuated tools left unloaded until they are ready to be used?
- [ ] Are power-actuated tools inspected for obstructions or defects each day before use?
- [ ] Do power-actuated tool operators have and use appropriate PPE such as hard hats, safety goggles, safety shoes and ear protectors?

MACHINE GUARDING

- [ ] Is there a training program to instruct employees on safe methods of machine operation?
- [ ] Is there adequate supervision to ensure that employees are following safe machine operating procedures?
- [ ] Is there a regular program of safety inspection of machinery and equipment?
- [ ] Is all machinery and equipment kept clean and properly maintained?
- [ ] Is sufficient clearance provided around and between machines to allow for safe operations, set up and servicing, material handling and waste removal?
- [ ] Is equipment and machinery securely placed and anchored to prevent tipping or other movement that could result in personal injury?
- [ ] Is there a power shut-off switch within reach of the operator's position at each machine?
- [ ] Can electric power to each machine be locked out for maintenance, repair, or security?
- [ ] Are the noncurrent-carrying metal parts of electrically operated machines bonded and grounded?
- [ ] Are foot-operated switches guarded or arranged to prevent accidental actuation by personnel or falling objects?
- [ ] Are manually operated valves and switches controlling the operation of equipment and machines clearly identified and readily accessible?
- [ ] Are all emergency stop buttons colored red?
- [ ] Are all pulleys and belts within 7 feet (2.1336 meters) of the floor or working level properly guarded?
- [ ] Are all moving chains and gears properly guarded?

- [ ] Are splash guards mounted on machines that use coolant to prevent the coolant from reaching employees?
- [ ] Are methods provided to protect the operator and other employees in the machine area from hazards created at the point of operation, ingoing nip points, rotating parts, flying chips and sparks?
- [ ] Are machine guards secure and arranged so they do not cause a hazard while in use?
- [ ] If special hand tools are used for placing and removing material, do they protect the operator's hands?
- [ ] Are revolving drums, barrels and containers guarded by an enclosure that is interlocked with the drive mechanism so that revolution cannot occur unless the guard enclosure is in place?
- [ ] Do arbors and mandrels have firm and secure bearings, and are they free from play?
- [ ] Are provisions made to prevent machines from automatically starting when power is restored after a power failure or shutdown?
- [ ] Are machines constructed so as to be free from excessive vibration when the largest size tool is mounted and run at full speed?
- [ ] If machinery is cleaned with compressed air, is air pressure controlled and PPE or other safeguards utilized to protect operators and other workers from eye and body injury?
- [ ] Are fan blades protected with a guard having openings no larger than l/2 inch (1.2700 centimeters) when operating within 7 feet (2.1336 meters) of the floor?
- [ ] Are saws used for ripping equipped with anti-kickback devices and spreaders?
- [ ] Are radial arm saws so arranged that the cutting head will gently return to the back of the table when released?

LOCKOUT/TAGOUT PROCEDURES

- [ ] Is all machinery or equipment capable of movement required to be de-energized or disengaged and blocked or locked out during cleaning, servicing, adjusting, or setting up operations?
- [ ] If the power disconnect for equipment does not also disconnect the electrical control circuit, are the appropriate electrical enclosures identified and is a means provided to ensure that the control circuit can also be disconnected and locked out?
- [ ] Is the locking out of control circuits instead of locking out main power disconnects prohibited?
- [ ] Are all equipment control valve handles provided with a means for locking out?
- [ ] Does the lockout procedure require that stored energy (mechanical, hydraulic, air, etc.) be released or blocked before equipment is locked out for repairs?
- [ ] Are appropriate employees provided with individually keyed personal safety locks?
- [ ] Are employees required to keep personal control of their key(s) while they have safety locks in use?
- [ ] Is it required that only the employee exposed to the hazard can place or remove the safety lock?
- [ ] Is it required that employees check the safety of the lockout by attempting a startup after making sure no one is exposed?
- [ ] Are employees instructed to always push the control circuit stop button prior to re-energizing the main power switch?
- [ ] Is there a means provided to identify any or all employees who are working on locked-out equipment by their locks or accompanying tags?
- [ ] Are a sufficient number of accident prevention signs or tags and safety padlocks provided for any reasonably foreseeable repair emergency?
- [ ] When machine operations, configuration, or size require an operator to leave the control station and part of the machine could move if accidentally activated, is the part required to be separately locked out or blocked?
- [ ] If equipment or lines cannot be shut down, locked out and tagged, is a safe job procedure established and rigidly followed?

WELDING, CUTTING AND BRAZING

- [ ] Are only authorized and trained personnel permitted to use welding, cutting, or brazing equipment?
- [ ] Does each operator have a copy of and follow the appropriate operating instructions?
- [ ] Are compressed gas cylinders regularly examined for obvious signs of defects, deep rusting, or leakage?
- [ ] Is care used in handling and storage of cylinders, safety valves, relief valves, etc., to prevent damage?
- [ ] Are precautions taken to prevent the mixture of air or oxygen with flammable gases, except at a burner or in a standard torch?
- [ ] Are only approved apparatuses (torches, regulators, pressure reducing valves, acetylene generators, manifolds) used?
- [ ] Are cylinders kept away from sources of heat and elevators, stairs, or gangways?
- [ ] Is it prohibited to use cylinders as rollers or supports?
- [ ] Are empty cylinders appropriately marked and their valves closed?
- [ ] Are signs posted reading "DANGER, NO SMOKING, MATCHES, OR OPEN LIGHTS," or the equivalent?
- [ ] Are cylinders, cylinder valves, couplings, regulators, hoses and apparatuses kept free of oily or greasy substances?
- [ ] Is care taken not to drop or strike cylinders?
- [ ] Are regulators removed and valve-protection caps put in place before moving cylinders, unless they are secured on special trucks?
- [ ] Do cylinders without fixed wheels have keys, handles, or non-adjustable wrenches on stem valves when in service?
- [ ] Are liquefied gases stored and shipped valve-end up with valve covers in place?
- [ ] Are employees trained never to crack a fuel gas cylinder valve near sources of ignition?
- [ ] Before a regulator is removed, is the valve closed and gas released?
- [ ] Is red used to identify the acetylene (and other fuel-gas) hose, green for the oxygen hose and black for inert gas and air hoses?
- [ ] Are pressure-reducing regulators used only for the gas and pressures for which they are intended?
- [ ] Is open circuit (no-load) voltage of arc welding and cutting machines as low as possible and not in excess of the recommended limits?
- [ ] Under wet conditions, are automatic controls for reducing no-load voltage used?
- [ ] Is grounding of the machine frame and safety ground connections of portable machines checked periodically?
- [ ] Are electrodes removed from the holders when not in use?
- [ ] Is it required that electric power to the welder be shut off when no one is in attendance?
- [ ] Is suitable fire extinguishing equipment available for immediate use?
- [ ] Is the welder forbidden to coil or loop welding electrode cable around his body?
- [ ] Are wet machines thoroughly dried and tested before use?
- [ ] Are work and electrode lead cables frequently inspected for wear and damage, and replaced when needed?
- [ ] Are cable connectors adequately insulated?
- [ ] When the object to be welded cannot be moved and fire hazards cannot be removed, are shields used to confine heat, sparks and slag?
- [ ] Are fire watchers assigned when welding or cutting is performed in locations where a serious fire might develop?
- [ ] Are combustible floors kept wet, covered with damp sand, or protected by fire-resistant shields?
- [ ] Are personnel protected from possible electrical shock when floors are wet?
- [ ] Are precautions taken to protect combustibles on the other side of metal walls when welding is underway?

- [ ] Are used drums, barrels, tanks and other containers thoroughly cleaned of substances that could explode, ignite, or produce toxic vapors before hot work begins?
- [ ] Do eye protection, helmets, hand shields and goggles meet appropriate standards?
- [ ] Are employees exposed to the hazards created by welding, cutting, or brazing operations protected with PPE and clothing?
- [ ] Is a check made for adequate ventilation in and where welding or cutting is performed?
- [ ] When working in confined places, are environmental monitoring tests done and means provided for quick removal of welders in case of an emergency?

COMPRESSORS AND COMPRESSED AIR

- [ ] Are compressors equipped with pressure relief valves and pressure gauges?
- [ ] Are compressor air intakes installed and equipped so as to ensure that only clean, uncontaminated air enters the compressor?
- [ ] Are air filters installed on the compressor intake?
- [ ] Are compressors operated and lubricated in accordance with the manufacturer's recommendations?
- [ ] Are safety devices on compressed air systems checked frequently?
- [ ] Before a compressor's pressure system is repaired, is the pressure bled off and the system locked out?
- [ ] Are signs posted to warn of the automatic starting feature of the compressors?
- [ ] Is the belt drive system totally enclosed to provide protection for the front, back, top and sides?
- [ ] Are employees strictly prohibited from directing compressed air towards a person?
- [ ] Are employees prohibited from using highly compressed air for cleaning purposes?
- [ ] When compressed air is used to clean clothing, are employees trained to reduce the pressure to less than 10 pounds per square inch (psi)?
- [ ] When using compressed air for cleaning, do employees wear protective chip guarding and PPE?
- [ ] Are safety chains or other suitable locking devices used at couplings of high-pressure hose lines where a connection failure would create a hazard?
- [ ] Before compressed air is used to empty containers of liquid, is the safe working pressure of the container checked?
- [ ] When compressed air is used with abrasive blast cleaning equipment, is the operating valve a type that must be held open manually?
- [ ] When compressed air is used to inflate auto tires, are a clip-on chuck and an inline regulator preset to 40 psi required?
- [ ] Are employees prohibited from using compressed air to clean up or move combustible dust if such action could cause the dust to be suspended in the air and cause a fire or explosion hazard?

COMPRESSORS/AIR RECEIVERS

- [ ] Is every receiver equipped with a pressure gauge and one or more automatic, spring-loaded safety valves?
- [ ] Is the total relieving capacity of the safety valve able to prevent pressure in the receiver from exceeding the maximum allowable working pressure of the receiver by more than 10 percent?
- [ ] Is every air receiver provided with a drain pipe and valve at the lowest point for the removal of accumulated oil and water?
- [ ] Are compressed air receivers periodically drained of moisture and oil?
- [ ] Are all safety valves tested at regular intervals to determine whether they are in good operating condition?
- [ ] Is there a current operating permit?
- [ ] Is the inlet of air receivers and piping systems kept free of accumulated oil and carbonaceous materials?

COMPRESSED GAS CYLINDERS

- ☐ Are cylinders with a water weight capacity over 30 pounds (13.6 kilograms) equipped with a means to connect a valve protector device, or with a collar or recess to protect the valve?
- ☐ Are cylinders legibly marked to clearly identify the type of gas?
- ☐ Are compressed gas cylinders stored in areas protected from external heat sources such as flame impingement, intense radiant heat, electric arcs, or high-temperature lines?
- ☐ Are cylinders located or stored in areas where they will not be damaged by passing or falling objects or subject to tampering by unauthorized persons?
- ☐ Are cylinders stored or transported in a manner to prevent them from creating a hazard by tipping, falling, or rolling?
- ☐ Are cylinders containing liquefied fuel gas stored or transported in a position so that the safety relief device is always in direct contact with the vapor space in the cylinder?
- ☐ Are valve protectors always placed on cylinders when the cylinders are not in use or connected for use?
- ☐ Are all valves closed off before a cylinder is moved, when the cylinder is empty and at the completion of each job?
- ☐ Are low-pressure fuel gas cylinders checked periodically for corrosion, general distortion, cracks, or any other defect that might indicate a weakness or render them unfit for service?
- ☐ Does the periodic check of low-pressure fuel gas cylinders include a close inspection of the cylinders' bottoms?

HOIST AND AUXILIARY EQUIPMENT

- ☐ Is each overhead electric hoist equipped with a limit device to stop the hook at its highest and lowest point of safe travel?
- ☐ Will each hoist automatically stop and hold any load up to 125 percent of its rated load if its actuating force is removed?
- ☐ Is the rated load of each hoist legibly marked and visible to the operator?
- ☐ Are stops provided at the safe limits of travel for trolley hoists?
- ☐ Are the controls of hoists plainly marked to indicate the direction of travel or motion?
- ☐ Is each cage-controlled hoist equipped with an effective warning device?
- ☐ Are close-fitting guards or other suitable devices installed on each hoist to ensure that hoist ropes will be maintained in the sheave grooves?
- ☐ Are all hoist chains or ropes long enough to handle the full range of movement of the application while maintaining two full wraps around the drum at all times?
- ☐ Are guards provided for nip points or contact points between hoist ropes and sheaves permanently located within 7 feet (2.1336 meters) of the floor, ground, or working platform?
- ☐ Are employees prohibited from using chains or rope slings that are kinked or twisted and prohibited from using the hoist rope or chain wrapped around the load as a substitute for a sling?
- ☐ Is the operator instructed to avoid carrying loads above people?

INDUSTRIAL TRUCKS - FORKLIFTS

- ☐ Are employees properly trained in the use of the type of industrial truck they operate?
- ☐ Are only trained personnel allowed to operate industrial trucks?
- ☐ Is substantial overhead protective equipment provided on high lift rider equipment?
- ☐ Are the required lift truck operating rules posted and enforced?
- ☐ Is directional lighting provided on each industrial truck that operates in an area with less than 2 footcandles per square foot of general lighting?
- ☐ Does each industrial truck have a warning horn, whistle, gong, or other device that can be clearly heard above normal noise in the areas where it is operated?

- ☐ Are the brakes on each industrial truck capable of bringing the vehicle to a complete and safe stop when fully loaded?
- ☐ Does the parking brake of the industrial truck prevent the vehicle from moving when unattended?
- ☐ Are industrial trucks that operate where flammable gases, vapors, combustible dust, or ignitable fibers may be present approved for such locations?
- ☐ Are motorized hand and hand/rider trucks designed so that the brakes are applied and power to the drive motor shuts off when the operator releases his or her grip on the device that controls the truck's travel?
- ☐ Are industrial trucks with internal combustion engines that are operated in buildings or enclosed areas carefully checked to ensure that such operations do not cause harmful concentrations of dangerous gases or fumes?
- ☐ Are safe distances maintained from the edges of elevated ramps and platforms?
- ☐ Are employees prohibited from standing or passing under elevated portions of trucks, whether loaded or empty?
- ☐ Are unauthorized employees prohibited from riding on trucks?
- ☐ Are operators prohibited from driving up to anyone standing in front of a fixed object?
- ☐ Are arms and legs kept inside the running lines of the truck?
- ☐ Are loads handled only within the rated capacity of the truck?
- ☐ Are trucks in need of repair removed from service immediately?

SPRAYING OPERATIONS

- ☐ Is adequate ventilation provided before spraying operations are started?
- ☐ Is mechanical ventilation provided when spraying operations are performed in enclosed areas?
- ☐ When mechanical ventilation is provided during spraying operations, is it so arranged that it will not circulate the contaminated air?
- ☐ Is the spray area free of hot surfaces and at least 20 feet (6.096 meters) from flames, sparks, operating electrical motors and other ignition sources?
- ☐ Are portable lamps used to illuminate spray areas suitable for use in a hazardous location?
- ☐ Is approved respiratory equipment provided and used when appropriate during spraying operations?
- ☐ Do solvents used for cleaning have a flash point to 100 degrees Fahrenheit (deg. F) or more?
- ☐ Are fire control sprinkler heads kept clean?
- ☐ Are "NO SMOKING" signs posted in spray areas, paint rooms, paint booths and paint storage areas?
- ☐ Is the spray area kept clean of combustible residue?
- ☐ Are spray booths constructed of metal, masonry, or other substantial noncombustible material?
- ☐ Are spray booth floors and baffles noncombustible and easily cleaned?
- ☐ Is infrared drying apparatus kept out of the spray area during spraying operations and is the spray booth completely ventilated before using the drying apparatus?
- ☐ Is the electric drying apparatus properly grounded?
- ☐ Are lighting fixtures for spray booths located outside the booth with the interior lighted through sealed clear panels?
- ☐ Are the electric motors for exhaust fans placed outside booths or ducts?
- ☐ Are belts and pulleys inside the booth fully enclosed?
- ☐ Do ducts have access doors to allow cleaning?
- ☐ Do all drying spaces have adequate ventilation?

ENTERING CONFINED SPACES

- ☐ Are confined spaces thoroughly emptied of any corrosive or hazardous substances, such as acids or caustics, before entry?

- ☐ Are all lines to a confined space that contain inert, toxic, flammable, or corrosive materials valved off and blanked or disconnected and separated before entry?
- ☐ Are all impellers, agitators, or other moving parts and equipment inside confined spaces locked out if they present a hazard?
- ☐ Is either natural or mechanical ventilation provided prior to confined space entry?
- ☐ Are appropriate atmospheric tests performed to check for oxygen deficiency, toxic substances and explosive concentrations in the confined space before entry?
- ☐ Is adequate illumination provided for the work to be performed in the confined space?
- ☐ Is the atmosphere inside the confined space frequently tested or continuously monitored during work?
- ☐ Is there a trained and equipped standby employee positioned outside the confined space, whose sole responsibility is to watch the work in progress, sound an alarm if necessary and render assistance?
- ☐ Is the standby employee appropriately trained and equipped to handle an emergency?
- ☐ Are employees prohibited from entering the confined space without lifelines and respiratory equipment if there is any question as to the cause of an emergency?
- ☐ Is approved respiratory equipment required if the atmosphere inside the confined space cannot be made acceptable?
- ☐ Is all portable electrical equipment used inside confined spaces either grounded and insulated or equipped with ground fault protection?
- ☐ Are compressed gas bottles forbidden inside the confined space?
- ☐ Before gas welding or burning is started in a confined space, are hoses checked for leaks, torches lighted only outside the confined area and the confined area tested for an explosive atmosphere each time before a lighted torch is taken into the confined space?
- ☐ If employees will be using oxygen-consuming equipment such as salamanders, torches, furnaces, etc., in a confined space, is sufficient air provided to assure combustion without reducing the oxygen concentration of the atmosphere below 19.5 percent by volume?
- ☐ Whenever combustion-type equipment is used in a confined space, are provisions made to ensure the exhaust gases are vented outside of the enclosure?
- ☐ Is each confined space checked for decaying vegetation or animal matter which may produce methane?
- ☐ Is the confined space checked for possible industrial waste which could contain toxic properties?
- ☐ If the confined space is below ground and near areas where motor vehicles will be operating, is it possible for vehicle exhaust or carbon monoxide to enter the space?

ENVIRONMENTAL CONTROLS

- ☐ Are all work areas properly illuminated?
- ☐ Are employees instructed in proper first aid and other emergency procedures?
- ☐ Are hazardous substances, blood and other potentially infectious materials, which may cause harm by inhalation, ingestion, or skin absorption or contact, identified?
- ☐ Are employees aware of the hazards involved with the various chemicals they may be exposed to in their work environment, such as ammonia, chlorine, epoxies, caustics, etc.?
- ☐ Is employee exposure to chemicals in the workplace kept within acceptable levels?
- ☐ Can a less harmful method or product be used?
- ☐ Is the work area ventilation system appropriate for the work performed?
- ☐ Are spray painting operations performed in spray rooms or booths equipped with an appropriate exhaust system?
- ☐ Is employee exposure to welding fumes controlled by ventilation, use of respirators, exposure time limits, or other means?

- ☐ Are welders and other nearby workers provided with flash shields during welding operations?
- ☐ If forklifts and other vehicles are used in buildings or other enclosed areas, are the carbon monoxide levels kept below maximum acceptable concentration?
- ☐ Has there been a determination that noise levels in the facilities are within acceptable levels?
- ☐ Are steps being taken to use engineering controls to reduce excessive noise levels?
- ☐ Are proper precautions being taken when handling asbestos and other fibrous materials?
- ☐ Are caution labels and signs used to warn of hazardous substances (e.g., asbestos) and biohazards (e.g., bloodborne pathogens)?
- ☐ Are wet methods used, when practicable, to prevent the emission of airborne asbestos fibers, silica dust and similar hazardous materials?
- ☐ Are engineering controls examined and maintained or replaced on a scheduled basis?
- ☐ Is vacuuming with appropriate equipment used whenever possible rather than blowing or sweeping dust?
- ☐ Are grinders, saws and other machines that produce respirable dusts vented to an industrial collector or central exhaust system?
- ☐ Are all local exhaust ventilation systems designed to provide sufficient air flow and volume for the application, and are ducts not plugged and belts not slipping?
- ☐ Is PPE provided, used and maintained wherever required?
- ☐ Are there written standard operating procedures for the selection and use of respirators where needed?
- ☐ Are restrooms and washrooms kept clean and sanitary?
- ☐ Is all water provided for drinking, washing and cooking potable?
- ☐ Are all outlets for water that is not suitable for drinking clearly identified?
- ☐ Are employees' physical capacities assessed before they are assigned to jobs requiring heavy work?
- ☐ Are employees instructed in the proper manner for lifting heavy objects?
- ☐ Where heat is a problem, have all fixed work areas been provided with spot cooling or air conditioning?
- ☐ Are employees screened before assignment to areas of high heat to determine if their health might make them more susceptible to having an adverse reaction?
- ☐ Are employees working on streets and roadways who are exposed to the hazards of traffic required to wear bright colored (traffic orange) warning vests?
- ☐ Are exhaust stacks and air intakes located so that nearby contaminated air will not be recirculated within a building or other enclosed area?
- ☐ Is equipment producing ultraviolet radiation properly shielded?
- ☐ Are universal precautions observed where occupational exposure to blood or other potentially infectious materials can occur and in all instances where differentiation of types of body fluids or potentially infectious materials is difficult or impossible?

FLAMMABLE AND COMBUSTIBLE MATERIALS

- ☐ Are combustible scrap, debris and waste materials (oily rags, etc.) stored in covered metal receptacles and promptly removed from the worksite?
- ☐ Is proper storage practiced to minimize the risk of fire, including spontaneous combustion?
- ☐ Are approved containers and tanks used to store and handle flammable and combustible liquids?
- ☐ Are all connections on drums and combustible liquid piping, vapor and liquid tight?
- ☐ Are all flammable liquids kept in closed containers when not in use (e.g., parts cleaning tanks, pans, etc.)?

- ☐ Are bulk drums of flammable liquids grounded and bonded to containers during dispensing?
- ☐ Do storage rooms for flammable and combustible liquids have explosion-proof lights and mechanical or gravity ventilation?
- ☐ Is liquefied petroleum gas stored, handled and used in accordance with safe practices and standards?
- ☐ Are "NO SMOKING" signs posted on liquefied petroleum gas tanks and in areas where flammable or combustible materials are used or stored?
- ☐ Are liquefied petroleum storage tanks guarded to prevent damage from vehicles?
- ☐ Are all solvent wastes and flammable liquids kept in fire-resistant, covered containers until they are removed from the worksite?
- ☐ Is vacuuming used whenever possible rather than blowing or sweeping combustible dust?
- ☐ Are firm separators placed between containers of combustibles or flammables that are stacked one upon another to ensure their support and stability?
- ☐ Are fuel gas cylinders and oxygen cylinders separated by distance and fire-resistant barriers while in storage?
- ☐ Are fire extinguishers selected and provided for the types of materials in the areas where they are to be used?

  Class A - Ordinary combustible material fires.

  Class B - Flammable liquid, gas or grease fires.

  Class C - Energized-electrical equipment fires.
- ☐ Are appropriate fire extinguishers mounted within 75 feet (22.86 meters) of outside areas containing flammable liquids and within 10 feet (3.048 meters) of any inside storage area for such materials?
- ☐ Are extinguishers free from obstructions or blockage?
- ☐ Are all extinguishers serviced, maintained and tagged at intervals not to exceed one year?
- ☐ Are all extinguishers fully charged and in their designated places?
- ☐ Where sprinkler systems are permanently installed, are the nozzle heads so directed or arranged that water will not be sprayed into operating electrical switchboards and equipment?
- ☐ Are safety cans used for dispensing flammable or combustible liquids at the point of use?
- ☐ Are all spills of flammable or combustible liquids cleaned up promptly?
- ☐ Are storage tanks adequately vented to prevent the development of excessive vacuum or pressure as a result of filling, emptying, or atmosphere temperature changes?
- ☐ Are storage tanks equipped with emergency venting that will relieve excessive internal pressure caused by fire exposure?
- ☐ Are rules enforced in areas involving storage and use of hazardous materials?

HAZARDOUS CHEMICAL EXPOSURE

- ☐ Are employees aware of the potential hazards and trained in safe handling practices for situations involving various chemicals stored or used in the workplace such as acids, bases, caustics, epoxies, phenols, etc.?
- ☐ Is employee exposure to chemicals kept within acceptable levels?
- ☐ Are eye-wash fountains and safety showers provided in areas where corrosive chemicals are handled?
- ☐ Are all containers, such as vats, storage tanks, etc., labeled as to their contents, e.g., "CAUSTICS"?
- ☐ Are all employees required to use personal protective clothing and equipment when handling chemicals (gloves, eye protection, respirators, etc.)?
- ☐ Are flammable or toxic chemicals kept in closed containers when not in use?
- ☐ Are chemical piping systems clearly marked as to their content?
- ☐ Where corrosive liquids are frequently handled in open containers or drawn from storage vessels or pipelines, are adequate means readily

available for neutralizing or disposing of spills or overflows and performed properly and safely?

- ☐ Are standard operating procedures established and are they being followed when cleaning up chemical spills?
- ☐ Are respirators stored in a convenient, clean and sanitary location, and are they adequate for emergencies?
- ☐ Are employees prohibited from eating in areas where hazardous chemicals are present?
- ☐ Is PPE used and maintained whenever necessary?
- ☐ Are there written standard operating procedures for the selection and use of respirators where needed?
- ☐ If you have a respirator protection program, are your employees instructed on the correct usage and limitations of the respirators? Are the respirators National Institute for Occupational Safety and Health (NIOSH)-approved for this particular application? Are they regularly inspected, cleaned, sanitized and maintained?
- ☐ If hazardous substances are used in your processes, do you have a medical or biological monitoring system in operation?
- ☐ Are you familiar with the threshold limit values or permissible exposure limits of airborne contaminants and physical agents used in your workplace?
- ☐ Have appropriate control procedures been instituted for hazardous materials, including safe handling practices and the use of respirators and ventilation systems?
- ☐ Whenever possible, are hazardous substances handled in properly designed and exhausted booths or similar locations?
- ☐ Do you use general dilution or local exhaust ventilation systems to control dusts, vapors, gases, fumes, smoke, solvents, or mists that may be generated in your workplace?
- ☐ Is operational ventilation equipment provided for removal of contaminants from production grinding, buffing, spray painting, and/or vapor degreasing?
- ☐ Do employees complain about dizziness, headaches, nausea, irritation, or other factors of discomfort when they use solvents or other chemicals?
- ☐ Is there a dermatitis problem? Do employees complain about dryness, irritation, or sensitization of the skin?
- ☐ Have you considered having an industrial hygienist or environmental health specialist evaluate your operation?
- ☐ If internal combustion engines are used, is carbon monoxide kept within acceptable levels?
- ☐ Is vacuuming used rather than blowing or sweeping dust whenever possible for cleanup?
- ☐ Are materials that give off toxic, asphyxiant, suffocating, or anesthetic fumes stored in remote or isolated locations when not in use?

HAZARDOUS SUBSTANCES COMMUNICATION

- ☐ Is there a list of hazardous substances used in your workplace and an MSDS readily available for each hazardous substance used?
- ☐ Is there a current written exposure control plan for occupational exposure to bloodborne pathogens and other potentially infectious materials, where applicable?
- ☐ Is there a written hazard communication program dealing with MSDSs, labeling and employee training?
- ☐ Is each container for a hazardous substance (i.e., vats, bottles, storage tanks, etc.) labeled with product identity and a hazard warning (communication of the specific health hazards and physical hazards)?
- ☐ Is there an employee training program for hazardous substances that includes:
  - an explanation of what an MSDS is and how to use and obtain one;
  - MSDS contents for each hazardous substance or class of substances;
  - explanation of "A Right to Know";

- identification of where an employee can see the written hazard communication program;
- location of physical and health hazards in particular work areas and the specific protective measures to be used; and
- details of the hazard communication program, including how to use the labeling system and MSDSs.

☐ Does the employee training program on the bloodborne pathogens standard contain the following elements:

- an accessible copy of the standard and an explanation of its contents;
- a general explanation of the epidemiology and symptoms of bloodborne diseases;
- an explanation of the modes of transmission of Bloodborne Pathogens;
- an explanation of the employer's exposure control plan and the means by which employees can obtain a copy of the written plan;
- an explanation of the appropriate methods for recognizing tasks and the other activities that may involve exposure to blood and other potentially infectious materials;
- an explanation of the use and limitations of methods that will prevent or reduce exposure, including appropriate engineering controls, work practices and PPE;
- information on the types, proper use, location, removal, handling, decontamination and disposal of PPE;
- an explanation of the basis for selection of PPE;
- information on the hepatitis B vaccine;
- information on the appropriate actions to take and persons to contact in an emergency involving blood or other potentially infectious materials;
- an explanation of the procedure to follow if an exposure incident occurs, including the methods of reporting the incident and the medical follow-up that will be made available;
- information on post-exposure evaluations and follow-up; and
- an explanation of signs, labels and color coding.

☐ Are employees trained in:

- how to recognize tasks that might result in occupational exposure;
- how to use work practice, engineering controls and PPE, and their limitations;
- how to obtain information on the types, selection, proper use, location, removal, handling, decontamination and disposal of PPE; and
- who to contact and what to do in an emergency.

ELECTRICAL

☐ Do you require compliance with OSHA standards for all contract electrical work?

☐ Are all employees required to report any obvious hazard to life or property in connection with electrical equipment or lines as soon as possible?

☐ Are employees instructed to make preliminary inspections and/or appropriate tests to determine conditions before starting work on electrical equipment or lines?

☐ When electrical equipment or lines are to be serviced, maintained, or adjusted, are necessary switches opened, locked out or tagged, whenever possible?

☐ Are portable electrical tools and equipment grounded or of the double insulated type?

☐ Are electrical appliances such as vacuum cleaners, polishers, vending machines, etc., grounded?

☐ Do extension cords have a grounding conductor?

☐ Are multiple plug adaptors prohibited?

☐ Are ground-fault circuit interrupters installed on each temporary 15 or 20 ampere, 120 volt alternating current (AC) circuit at locations where construction, demolition, modifications, alterations, or excavations are being performed?

- ☐ Are all temporary circuits protected by suitable disconnecting switches or plug connectors at the junction with permanent wiring?
- ☐ Do you have electrical installations in hazardous dust or vapor areas? If so, do they meet the National Electrical Code (NEC) for hazardous locations?
- ☐ Are exposed wiring and cords with frayed or deteriorated insulation repaired or replaced promptly?
- ☐ Are flexible cords and cables free of splices or taps?
- ☐ Are clamps or other securing means provided on flexible cords or cables at plugs, receptacles, tools, equipment, etc., and is the cord jacket securely held in place?
- ☐ Are all cord, cable and raceway connections intact and secure?
- ☐ In wet or damp locations, are electrical tools and equipment appropriate for the use or location or otherwise protected?
- ☐ Is the location of electrical power lines and cables (overhead, underground, under floor, other side of walls, etc.) determined before digging, drilling, or similar work is begun?
- ☐ Are metal measuring tapes, ropes, hand-lines or similar devices with metallic thread woven into the fabric prohibited where they could come in contact with energized parts of equipment or circuit conductors?
- ☐ Is the use of metal ladders prohibited where the ladder or the person using the ladder could come in contact with energized parts of equipment, fixtures, or circuit conductors?
- ☐ Are all disconnecting switches and circuit breakers labeled to indicate their use or equipment served?
- ☐ Are disconnecting means always opened before fuses are replaced?
- ☐ Do all interior wiring systems include provisions for grounding metal parts of electrical raceways, equipment and enclosures?
- ☐ Are all electrical raceways and enclosures securely fastened in place?
- ☐ Are all energized parts of electrical circuits and equipment guarded against accidental contact by approved cabinets or enclosures?
- ☐ Is sufficient access and working space provided and maintained around all electrical equipment to permit ready and safe operations and maintenance?
- ☐ Are all unused openings (including conduit knockouts) in electrical enclosures and fittings closed with appropriate covers, plugs, or plates?
- ☐ Are electrical enclosures such as switches, receptacles, junction boxes, etc., provided with tight-fitting covers or plates?
- ☐ Are disconnecting switches for electrical motors in excess of two horsepower able to open the circuit when the motor is stalled without exploding? (Switches must be horsepower rated equal to or in excess of the motor rating.)
- ☐ Is low voltage protection provided in the control device of motors driving machines or equipment that could cause injury from inadvertent starting?
- ☐ Is each motor disconnecting switch or circuit breaker located within sight of the motor control device?
- ☐ Is each motor located within sight of its controller or is the controller disconnecting means able to be locked open or is a separate disconnecting means installed in the circuit within sight of the motor?
- ☐ Is the controller for each motor that exceeds two horsepower rated equal to or above the rating of the motor it serves?
- ☐ Are employees who regularly work on or around energized electrical equipment or lines instructed in cardiopulmonary resuscitation (CPR)?
- ☐ Are employees prohibited from working alone on energized lines or equipment over 600 volts?

NOISE

- [ ] Are there areas in the workplace where continuous noise levels exceed 85 decibels?
- [ ] Is there an ongoing preventive health program to educate employees in safe levels of noise, exposures, effects of noise on their health and the use of personal protection?
- [ ] Have work areas where noise levels make voice communication between employees difficult been identified and posted?
- [ ] Are noise levels measured with a sound level meter or an octave band analyzer and are records being kept?
- [ ] Have engineering controls been used to reduce excessive noise levels? Where engineering controls are determined to be infeasible, are administrative controls (i.e., worker rotation) being used to minimize individual employee exposure to noise?
- [ ] Is approved hearing protective equipment (noise attenuating devices) available to every employee working in noisy areas?
- [ ] Have you tried isolating noisy machinery from the rest of your operation?
- [ ] If you use ear protectors, are employees properly fitted and instructed in their use?
- [ ] Are employees in high noise areas given periodic audiometric testing to ensure that you have an effective hearing protection system?

FUELING

- [ ] Are employees prohibited from fueling an internal combustion engine with a flammable liquid while the engine is running?
- [ ] Are fueling operations performed to minimize spillage?
- [ ] When spillage occurs during fueling operations, is the spilled fuel washed away completely, evaporated, or are other measures taken to control vapors before restarting the engine?
- [ ] Are fuel tank caps replaced and secured before starting the engine?
- [ ] In fueling operations, is there always metal contact between the container and the fuel tank?
- [ ] Are fueling hoses designed to handle the specific type of fuel?
- [ ] Are employees prohibited from handling or transferring gasoline in open containers?
- [ ] Are open lights, open flames, sparking, or arcing equipment prohibited near fueling or transfer of fuel operations?
- [ ] Is smoking prohibited in the vicinity of fueling operations?
- [ ] Are fueling operations prohibited in buildings or other enclosed areas that are not specifically ventilated for this purpose?
- [ ] Where fueling or transfer of fuel is done through a gravity flow system, are the nozzles self-closing?

IDENTIFICATION OF PIPING SYSTEMS

- [ ] When nonpotable water is piped through a facility, are outlets or taps posted to alert employees that the water is unsafe and not to be used for drinking, washing, or other personal use?
- [ ] When hazardous substances are transported through above-ground piping, is each pipeline identified at points where confusion could introduce hazards to employees?
- [ ] When pipelines are identified by color painted bands or tapes, are the bands or tapes located at reasonable intervals and at each outlet, valve, or connection, and are all visible parts of the line so identified?
- [ ] When pipelines are identified by color, is the color code posted at all locations where confusion could introduce hazards to employees?
- [ ] When the contents of pipelines are identified by name or name abbreviation, is the information readily visible on the pipe near each valve or outlet?
- [ ] When pipelines carrying hazardous substances are identified by tags, are the tags constructed of durable materials, the message printed

clearly and permanently, and are tags installed at each valve or outlet?

- ☐ When pipelines are heated by electricity, steam, or other external source, are suitable warning signs or tags placed at unions, valves, or other serviceable parts of the system?

MATERIALS HANDLING

- ☐ Is there safe clearance for equipment through aisles and doorways?
- ☐ Are aisleways permanently marked and kept clear to allow unhindered passage?
- ☐ Are motorized vehicles and mechanized equipment inspected daily or prior to use?
- ☐ Are vehicles shut off and brakes set prior to loading or unloading?
- ☐ Are containers of liquid combustibles or flammables, when stacked while being moved, always protected by dunnage (packing material) sufficient to provide stability?
- ☐ Are dock boards (bridge plates) used when loading or unloading operations are taking place between vehicles and docks?
- ☐ Are trucks and trailers secured from movement during loading and unloading operations?
- ☐ Are dock plates and loading ramps constructed and maintained with sufficient strength to support imposed loading?
- ☐ Are hand trucks maintained in safe operating condition?
- ☐ Are chutes equipped with sideboards of sufficient height to prevent the materials being handled from falling off?
- ☐ Are chutes and gravity roller sections firmly placed or secured to prevent displacement?
- ☐ Are provisions made to brake the movement of the handled materials at the delivery end of rollers or chutes?
- ☐ Are pallets usually inspected before being loaded or moved?
- ☐ Are safety latches and other devices being used to prevent slippage of materials off of hoisting hooks?
- ☐ Are securing chains, ropes, chockers, or slings adequate for the job?
- ☐ Are provisions made to ensure that no one is below when hoisting material or equipment?
- ☐ Are MSDSs available to employees handling hazardous substances?

TRANSPORTING EMPLOYEES AND MATERIALS

- ☐ Do employees who operate vehicles on public thoroughfares have valid operator's licenses?
- ☐ When seven or more employees are regularly transported in a van, bus, or truck, is the operator's license appropriate for the class of vehicle being driven and are there enough seats?
- ☐ Are vehicles used to transport employees equipped with lamps, brakes, horns, mirrors, windshields and turn signals, and are they in good repair?
- ☐ Are transport vehicles provided with handrails, steps, stirrups, or similar devices, placed and arranged to allow employees to safely mount or dismount?
- ☐ Are employee transport vehicles equipped at all times with at least two reflective-type flares?
- ☐ Is a fully charged fire extinguisher, in good condition, with at least a 4 B:C rating maintained in each employee transport vehicle?
- ☐ When cutting tools or tools with sharp edges are carried in passenger compartments of employee transport vehicles, are they placed in closed boxes or containers that are secured in place?
- ☐ Are employees prohibited from riding on top of any load that could shift, topple, or otherwise become unstable?

CONTROL OF HARMFUL SUBSTANCES BY VENTILATION

- ☐ Is the volume and velocity of air in each exhaust system sufficient to gather the dusts, fumes, mists, vapors, or gases to be controlled, and to convey them to a suitable point of disposal?

- [ ] Are exhaust inlets, ducts and plenums designed, constructed and supported to prevent collapse or failure of any part of the system?
- [ ] Are clean-out ports or doors provided at intervals not to exceed 12 feet (3.6576 meters) in all horizontal runs of exhaust ducts?
- [ ] Where two or more different operations are being controlled through the same exhaust system, could the combination of substances involved create a fire, explosion, or chemical reaction hazard in the duct?
- [ ] Is adequate makeup air provided to areas where exhaust systems are operating?
- [ ] Is the source point for makeup air located so that only clean, fresh air, free of contaminants will enter the work environment?
- [ ] Where two or more ventilation systems serve a work area, is their operation such that one will not offset the functions of the other?

SANITIZING EQUIPMENT AND CLOTHING

- [ ] Is required personal protective clothing or equipment able to be cleaned and disinfected easily?
- [ ] Are employees prohibited from interchanging personal protective clothing or equipment, unless it has been properly cleaned?
- [ ] Are machines and equipment that process, handle, or apply materials that could injure employees cleaned and/or decontaminated before being overhauled or placed in storage?
- [ ] Are employees prohibited from smoking or eating in any area where contaminants are present that could be injurious if ingested?
- [ ] When employees are required to change from street clothing into protective clothing, is a clean change room with a separate storage facility for street and protective clothing provided?
- [ ] Are employees required to shower and wash their hair as soon as possible after a known contact with a carcinogen has occurred?
- [ ] When equipment, materials, or other items are taken into or removed from a carcinogen-regulated area, is it done in a manner that will not contaminate non-regulated areas or the external environment?

TIRE INFLATION

- [ ] Where tires are mounted and/or inflated on drop center wheels or on wheels with split rims and/or retainer rings, is a safe practice procedure posted and enforced?
- [ ] Does each tire inflation hose have a clip-on chuck with at least 2.54 inches (6.45 centimeters) of hose between the chuck and an in-line hand valve and gauge?
- [ ] Does the tire inflation control valve automatically shut off the air flow when the valve is released?
- [ ] Is a tire restraining device such as a cage, rack, or other effective means used while inflating tires mounted on split rims or rims using retainer rings?
- [ ] Are employees prohibited from standing directly over or in front of a tire while it is being inflated?

Index

O

P

R